Australian
FROGS

A Natural History

Australian
FROGS
A Natural History

Michael J. Tyler

Cornell University Press

ITHACA AND LONDON

First published in 1998 by Cornell University Press
Sage House, 512 East State Street,
Ithaca, New York, 14850

Librarian: Library of Congress
Cataloguing-in-Publication Data are available.

Printed and bound in Singapore.

ISBN 0-8014-8499-5

CONTENTS

ACKNOWLEDGMENTS

The study of the lives of frogs is largely a nocturnal activity and usually a team effort. Therefore, I would like to thank first of all, those who have been my field companions. In particular I owe a debt of gratitude to Margaret Davies, Angus Martin and Graeme Watson. Undertaking field studies with them has been a memorable experience and one I have enjoyed enormously. Much of the data that I have included in this book has been obtained in their company.

The various field studies were made possible by grants from the Australian Research Grants Scheme, the Supervising Scientist for the Alligator Rivers Region, the University of Adelaide, Australian Geographic and the Utah Foundation. Further logistic support was provided by the Northern Australia Research Unit of the Australian National University, the Department of Agriculture, WA, and by Ansett Airlines. Licences to undertake research on frogs were issued by the Conservation Commission of the Northern Territory, and by the Department of Conservation and Land Management of Western Australia.

The line drawings were expertly executed by Ruth Evans and Margaret Davies; the scanning electron micrographs were prepared by Chris Miller, Kerstin Lungershausen and Veronica Ward.

I am also indebted to colleagues who checked the entries in various tables and drew my attention to additional references in the literature; these were Mike Cappo, Bill Freeland, Glen Ingram, Murray Littlejohn, Angus Martin, Keith McDonald, Helen Panter, Dale Roberts and Graeme Watson. Any errors or omissions are my responsibility. Margaret Davies also provided constructive comments on the penultimate draft of the manuscript, whilst Glen Ingram, Keith McDonald and Ross Sadlier assisted in the preparation of the Appendix.

Some of the illustrations were re-drawn from figures appearing in journals. For permission to include them in this book I thank the American Society of Ichthyology and Herpetology *(Copeia)*; the Bureau of Meterology, Melbourne; the CSIRO *(Australian Journal of Wildlife Research)*; the Society for the Study of Amphibians and Reptiles *(Journal of Herpetology)*; the Royal Society of South Australia *(Transactions of the Royal Society of South Australia)*; Academic Press, London *(Biological Journal of the Linnean Society)*; the Queensland Museum *(Memoirs of the Queensland Museum)*; Elsevier Science Publishing Co., New York *(Gastroenterology)*; and the Royal Zoological Society *(Australian Zoologist)*. For the black and white photographs I would like to thank M. Davies (Figures 2, 51, 52, 53, 56); P. Kempster (Figures 5, 16, 20, 39, 57, 58); V. Ward (Figure 10); C. Miller (Figures 13, 17, 18, 38, 41, 49, 50); G.J. Ingram (Figure 55); R. McDonald (Figure 60); and D. Caville (Figure 61).

The manuscript was not simply typed, but also vetted for inconsistences of style and edited by Lorna Lucas, Sandra Lawson and Sue Dyer. My thanks to them all.

Finally, I thank the University of Adelaide for providing research facilities, and its Department of Zoology for creating an academic environment in which fundamental research into the nature and biology of the Australian fauna is encouraged.

COLOUR PLATES

TABLES

PREFACE

T his is the third occasion when I have had the opportunity to produce a book about the natural history of Australia's frogs. Essentially this one involves an update of *Australian Frogs* published in 1989. The manuscript was completed in 1987 and the book has been long out of print.

At a time when there is enormous interest in Australian frogs, I believe that it is highly desirable that a book on the ways and needs of these animals should be available. Is the interest in frogs really so substantial or is Tyler providing a biased view? I think I can confirm my opinion because of the response to evidence that many frog species are in serious decline, and that two or three species have possibly become extinct. The interest in frogs by naturalists, and folk who wouldn't claim to be naturalists, has increased as a consequence.

'Frogwatch' programs involving monitoring of populations by school children have been undertaken in all States and in the Northern Territory. Field guides have been published for the first time for Victoria and for south-eastern Australia, whilst one for Western Australia has been revised. Harold Cogger's field guide to all Australian species has been revised twice and a new field guide by John Barker, Gordon Grigg and myself is in press.

It is against this background that *Australian Frogs* has been revised. The original objective of it being a resource has been retained, so most of the tables have been updated to include references to works published since 1987. Some chapters are virtually unaltered, implying that there has been no significant change in the state of knowledge. In others there is supplementary text on topics such as the fossil record (so vital to understanding the historical background), and the cane toad, *Bufo marinus*, where there has been unprecedented study seeking a method of controlling populations.

The topic of declining frogs has become the subject of a new chapter, and I have included for ready reference a complete list of State and Territory field guides. Hopefully this edition is an improvement upon the first.

Michael J. Tyler
Adelaide, April 1994

CHAPTER ONE

ORIGINS

ORIGIN OF FROGS

The word 'Amphibia' is derived from the Greek words 'amphi' meaning double, and 'bios' meaning life. Thus classically it describes creatures which in the course of their life inhabit two different environments: first, water as eggs and tadpoles, and latterly, land or elsewhere out of water. Unfortunately this definition is inaccurate for a significant number of species which, in fact, do not lay eggs in water, but instead spend their life entirely upon land and lay their eggs there. A few more are totally aquatic as adults.

The origins of the modern Amphibia, such as frogs, has been the subject of considerable debate, a fact indicating the difficulty involved in tracing their ancestry back to the early Amphibia that roamed the earth, and also to the fish stocks from which they, in turn, had evolved.

On one point there seems to be agreement, namely that the ancestor of the Amphibia was a bony fish of the Class Osteichthyes which was widespread at the period when the first Amphibia emerged. But just which kind of osteichthyan produced the basic amphibian stock is disputed hotly. Some scientists favour the Dipnoi, better known as lung-fishes, of which one member *(Neoceratodus leichardti)* lives in Australia today. Others favour a second group called the Actinopterygii.

Irrespective of which group provided the source for the Amphibia, it is known that the first Amphibia made their appearance towards the end of the Devonian period approximately 345 million years ago.

The most primitive of the early Amphibia were creatures called Labyrinthodonts, named after the peculiarly labyrinth-like appearance of a transverse section of their teeth. These were large, long-bodied, short-limbed creatures resembling a crocodile with fins upon its tail and measuring up to 4 m in total length.

Just when and how the first frogs evolved remains unknown. The most significant fossil is accepted to be a Madagascan (Malagasy) creature named *Triadobatrachus* that lived 200 million years ago, but whether it can be considered a real frog or perhaps a pre-frog is yet another example of the difference in opinions held by zoologists. There is one feature which makes its classification as a frog rather problematical, namely that it has a more elongate body with no less than 16 vertebrae in front of the sacrum vertebra. This compares with a maximum of nine in modern forms of frogs. Irrespective of this difference (and of the significance attributed to it) *Triadobatrachus* has a number of characteristics that are frog-like.

RELATIONSHIPS OF MODERN AMPHIBIA

Today there are three separate groups of Amphibia in existence: Anura (frogs), Gymnophiona (legless, worm-like creatures) and the Caudata (formerly called the Urodela) which include newts and salamanders that possess limbs and a tail as well. The Anura is the only group to occur naturally in Australia. Newts have been introduced from time to time, the earliest incident probably that of an animal dealer who, in 1904, imported the fire-bellied newt, *Cynops* (as

Triturus) pyrrhogaster, from China (Waite, 1904). The Gymnophiona has a patchy distribution across Africa, Asia and South America. Its distribution approaches Australia, with one species in Sabah and a second species common to Sabah, Sarawak and the Philippines (Inger, 1954, 1966).

Whether these three groups (Orders) are derived from a common ancestor, or whether they evolved from two or more separate stocks remains uncertain. For a review of the anatomical evidence a good source of information are the accounts of Duellman and Trueb (1986) and certainly I think it is fair to say that most zoologists favour a single origin with the result that the groups are united together as shown in the following classification:

CLASS:	AMPHIBIA
Subclass:	Labyrinthodontia
Subclass:	Lissamphibia
Order:	Anura
Order:	Gymnophiona
Order:	Caudata

What is so fascinating is that the three modern orders of Amphibia are so different from one another. For example, members of the Gymnophiona (or caecilians as they are commonly called) have snake-like bodies, scales and vestigial eyes, whilst the Caudata include some creatures with two pairs of limbs and others (known as sirens) with only one pair. Most Caudata are less than 150 mm from the tip of the nose to the end of the tail, but the largest *(Andrias davidianus)* can reach a total length of 1.5 m. The degree of morphological divergence between the living orders is vast, implying that since their common origin they have followed totally different evolutionary paths. Nevertheless there are many similar examples of extreme divergence within classes of the animal kingdom—for example, the difference between various orders of mammals as represented by the mouse, giraffe and humans.

FROG FOSSILS

The earliest known frog fossil that really looks like a frog is *Vieraella herbstii* of Argentina which is approximately 160 million years old. It is about 28 mm long and has nine presacral vertebrae. Its close affinity to existing

Plate 1 *Neville Pledge screen washing sediments containing fossils at Lake Palankarinna, SA. (South Australian Museum)*

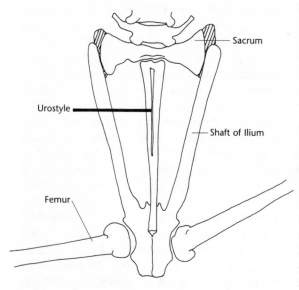

FIGURE 1 *Position of the ilium bones in the pelvis.*

species of frogs suggests that the first frogs evolved a good deal earlier.

It is generally accepted that the Australian frog fauna is derived from two sources separated by a wide interval of time. The first of these is the Gondwanan stock which would have occupied the southern continents prior to the fragmentation of the land mass. The second stock is most likely of an Oriental source which reached Australia, probably via New Guinea, after the Australian continental plate had collided with the Oriental plate in the Mid-Miocene, some 15 million years ago.

The Gondwanan stock is represented by the families Hylidae and Leptodactylidae, whilst the Oriental stock includes the families Microhylidae and Ranidae. Not surprisingly the Gondwanan stock occurs throughout the Australian continent, whilst the Oriental has not penetrated far from its northern entry point and now, with the exception of *Sphenophryne adelphe* and *Rana daemeli* in the Northern Territory, is confined to the Cape York Peninsula of northern Queensland.

FOSSILS IN AUSTRALIA

At present no intact fossil frog has been found in Australia. In fact all finds to date have been of isolated, disarticulated bones. What progress has been made in establishing the fossil record has occurred entirely since 1974 when the first specimen was reported from the dry bed of Lake Palankarinna north of Lake Eyre in South Australia. We now know of over 2000.

Identifying species or even genera of frogs from single bones is a difficult proposition. In a few isolated cases there are unique features which permit identification, and by far the most revealing of all of the bones of the body is the ilium (plural = ilia). The ilium is the largest bone of the pelvis. It is a paired structure united at the base where it provides a socket for the long femur bone of the hindleg. It then passes forward in a long shaft and is attached on each side by ligaments to the outer edges of the sacral vertebra (Figure 1). Study of the ilium is instructive for a number of reasons. First, because it is possible to estimate the size of the frog involved by the relative length of the ilium and, secondly, because the shape, size and position of various processes on the bone provide an indication of a frog's habits. For example, species that burrow tend to have an enlarged zone around the articulation of the femur, providing increased sites for attachment of the large muscles of the strong legs.

Locating ilia in fossil deposits is a relatively simple, but time-consuming, process. If a sufficient portion of the ilial shaft is complete the bone tends to be curved and so will not lie flat upon a sorting tray. It is easy to

FIGURE 2 *Head of an ilium bone from the pelvis of* Litoria ewingi. *A fossil from a deposit at Henschke's Cave, SA, perhaps 20 000 years old.*

0.5 mm

FIGURE 3 *Sorting bones excavated from a fossil deposit.*

form a 'search-image' for such bones when sorting. The usual method is simply to take a cupful of mixed bones, dust and matrix from a fossil deposit, spread it out upon a white plastic tray and, with the aid of a soft, camel hair brush, separate to one side each frog bone (Figure 2). It is possible to become sufficiently skilled after a while so that each kind of bone there will be recognised (Figure 3) and, at speed, the only bone likely to be momentarily confused with an ilium is a mammalian rib because it too is curved.

Many of the fossil bones found to date have not resulted from a specific field search for frogs. Rather, the fossil hunters have been searching for marsupial remains (which are more popular), and the bones involved tend to be large. To free these bones from the surrounding soil matrix they are washed upon screen sieves. The fragile frog bones included are found at a later stage amongst the debris, and unfortunately their passage across the screens means that most become broken in the process. But broken bones are better than no bones at all (Plate 1).

At Riversleigh Station in north-west Queensland the fossils are encased in solid limestone and are extracted by dissolving the rock in vats of acetic acid.

A complete list of the fossil frog species recovered so far is shown in Table 1 and the sites from which they have been found are shown in Figure 4. To date 32 species have been located at 14 sites. The youngest material is from Hunter Island off the coast of Tasmania and is less than 10 000 years old (Holocene). The oldest is from a clay deposit at Murgon, Boat Mountain in south-east Queensland and has been dated at 54.5 million years. This material therefore antedates the separation of Australia from Antarctica linking the continent to South America.

What is so surprising about the Murgon frog fauna is that it consists predominantly of genera that exist in Australia today. Their evolution clearly was not a consequence of the period of geographic isolation when Australia drifted north away from Antarctica; for some reason Australia acquired its distinctive frog fauna during an earlier period.

Caves feature conspicuously amongst the sites listed in Table 1 and without exception the deposits are in the vicinity of 50 000 years old (late Pleistocene). Most of these deposits

TABLE ONE: Fossil frogs of Australia

GENUS & SPECIES	LOCALITY	AGE OF DEPOSIT	REFERENCE
Australobatrachus ilius	L. Palankarinna, SA	Oligocene	Tyler (1974, 1976a, 1982, 1986)
" "	L. Yanda, SA	Mid-Miocene	Tyler (1986)
Australobatrachus undulatus	Bullock Creek	Miocene	Tyler (1994)
Crinia georgiana	Skull Cave, WA	Pleistocene	Tyler (1985a)
" "	Devil's Lair, WA	"	Tyler (1985a)
Crinia presignifera	Riversleigh Stn, Qld	Oligo-Miocene	Tyler (1991a)
Crinia remota	Riversleigh Stn, Qld	Holocene-Late Pleistocene	Tyler (1991a)
Crinia signifera	Henschke's Cave, SA	"	Tyler (1977)
" "	Victoria Cave, SA	"	Tyler (1977)
Cyclorana cultripes	Floraville Station	Plio-Pleistocene	Tyler, Godthelp & Archer (1994)
Cyclorana platycephala	Floraville Station	Plio-Pleistocene	Tyler, Godthelp & Archer (1994)
Geocrinia sp. ? *laevis*	Victoria Cave, SA	Plio-Pleistocene	Tyler (1977)
Heleioporus/Neobatrachus spp.	Skull Cave, WA	Plio-Pleistocene	Tyler (1985)
Kyarranus borealis	Riversleigh Stn, Qld	Oligo-Miocene	Tyler (1991b)
Lechriodus casca	Murgon, Qld	Early Eocene	Tyler & Godthelp (1993)
Lechriodus intergerivus	Riversleigh Stn, Qld	Oligo-Miocene	Tyler (1989); Tyler, Hand & Ward (1990)
Limnodynastes antecessor	Riversleigh Stn, Qld	Early Miocene	Tyler (1989)
Limnodynastes archeri	L. Palankarinna, SA	Oligocene	Tyler (1982)
Limnodynastes dorsalis	Skull Cave, WA	Pleistocene	Tyler (1985)
Limnodynastes dumerilii	L. Menindee, NSW	Pleistocene	Tyler (unpublished)
Limnodynastes sp. nr *dumerilii*	Victoria Cave, SA	Pleistocene	Tyler (1985)
Limnodynastes sp. nr *peronii*	Hunter Island, Tas.	Holocene	Tyler (unpublished)
Limnodynastes tasmaniensis	Henschke's Cave, SA	Pleistocene	Tyler (1977)
" "	Victoria Cave, SA	"	Tyler (1977)
Limnodynastes cf. *tasmaniensis*	Floraville Station	Plio-Pleistocene	Tyler, Godthelp & Archer (1994)
Litoria adelaidensis	Skull Cave, WA	"	Tyler (1985)
" "	Devil's Lair, WA	"	Tyler (1985)
Litoria conicula	Bullock Creek, NT	Miocene	Tyler (1994)
Litoria curvata	Bullock Creek, NT	Miocene	Tyler (1994)
Litoria ewingii	Henschke's Cave, SA	Pleistocene	Tyler (1977)
" "	Victoria Cave, SA	"	Tyler (1977)
Litoria magna	Riversleigh Stn, Qld	Oligo-Miocene	Tyler (1991c)
Litoria sp.	L. Palankarinna, SA	Oligocene	Tyler (1982)
Litoria sp.	L. Palankarinna, SA	"	Estes (1984)
Litoria sp. ? *caerulea*	L. Palankarinna, SA	Oligocene	Tyler (1985)
Litoria sp. nr *cyclorhynchus* & *moorei*	Skull Cave, WA	Pleistocene	Tyler (1985)
Neobatrachus pictus	Curramulka, SA	Pliocene	Tyler (1988)
Pseudophryne guentheri	Devil's Lair, WA	Pleistocene	Tyler (1985)

FIGURE 4 *Sites in Australia where frog fossils have been found.*

resulted from the activity of owls which disgorged pellets of indigestible material including bones of animals which formed part of their prey. In contrast the *Crinia remota* from Carrington Cave at Riversleigh Station was found amongst a vast heap of bone fragments dropped by bats from the roof above.

At present no fossils have been located which give any indication of the assumed evolutionary links with the fauna of South America. The most likely source of material to resolve this tantalising issue will be the finding of a freshwater deposit of Cretaceous age (65–135 million years).

FROGS IN ROCK

Australia may be without any intact frog fossils but there are a number of reports of living frogs being found in solid rock. It is difficult to take most of these reports seriously, and it may seem on the fringe of heresy to include an account of these phenomena in a factual discussion of the history of Australian frogs. That I do so reflects a nagging doubt in my mind. This doubt exists because, whereas some accounts border on the ludicrous, there remain a few which have a ring of authenticity to them.

A few years ago I broadcast a talk on frogs for the ABC Science Show. I gave an account of the known longevity of frogs, of experiments performed to determine how long

frogs could live in rock, and cited letters I had received from folk who claimed to have seen frogs stepping out from cavities in rock.

After the broadcast I received more letters and phone calls from people who had heard of similar phenomena or who had witnessed other frogs in rock.

Take for example the graphic account of Mr Martin Cash (then aged 93) of an incident that took place at Kalka Station near Streaky Bay, South Australia.

Yes that is right! I did find a toad or frog-like creature inside a stone. I am an old man now and those I had called as witness as well as those I had told of it in conversation are all now dead and gone. I remembered the incident again as I lay awake in the early morning of today, and telling the story at the breakfast table a relative persuaded me to write a short account of the happening.

It was Spring time of the year around 1916 or '17, that I was at some task in the old smithy at Kalka Station where I was reared. The old smithy, built some 50 years before, was in bad shape. Some stones had been dislodged from the front corner of the wide open doorway and could have been subjected to drip from the roof which never had a spouting. I cannot now remember what task I was about but I had a 2 lb hammer in my hand and just why I should have hit one of the old stones I cannot now say but I gave the stone one heavy blow of the hammer and it split in two showing two clean faces although blackened elsewhere. I was astounded to see a live toad or frog encased in a cavity into which it exactly fitted. This cavity was mostly on one side of the break line and it seems that in breaking, the two sides of the broken stone slid tightly past the cavity instead of falling straight apart.

The poor creature was fatally injured, having one side of its body torn open. However it made no sound but I did see it move a leg although its inside part was exposed. It was a very light green with yellow in parts and I cannot recall anything about its eyes but that is no proof that it did not have eyes though they do not seem to have been staring at me. Its broken body exposed no organs that I could see but only a thick pale-yellow fluid too thick to flow. It did not bleed. I did not have the heart to kill it so I laid it gently on a small earth heap some 20 yards away. I called a younger brother to look at the toad and the cavity in the stone. So we left it and went to the house for morning tea but when we returned we looked for it again; but it was gone. Recovery seemed impossible and some seagulls 30 yards away were blamed for its disappearance.

The stone in which the toad was encased was a porous inferior limestone, really calcareous sandstone, and the cavity was larger than a ping-pong ball. There was no vein or tree root-track in the stone where some embryo may more easily have entered, but that must have happened, and what sustained life is hard to imagine with only air, moisture, lime and silica available. Toads all alive but encased in stones have been reported in various parts of the world and for a few to be found alive it must occur not infrequently for I feel sure that not one in 10 000 is broken open while the toad is living. A question that has some bearing on their life span in the stone is this: Did anyone ever find a dead 'Toad in the Stone'?

I've come across two dead frogs in stone, one is exhibited in a glass case in a gallery at the museum in Brighton, England. There, within one of the antiquated exhibition cases, is a large flint split into two portions. The split divides the stone and separates a large central cavity into two. In one half a small tunnel, partly blocked with chalk, communicates between the centre and the outside. The flint was brought to the museum by Charles Dawson who wrote an account of his find in the Brighton Museum Journal. He argued that a small toad had entered the cavity through the aperture and that insects would have been attracted because of the smell of the toad. As the toad ate the insects so it grew until it reached the stage of being unable to escape through the entrance. Eventually it died, the entrance became plugged with chalk and the toad dehydrated in the dry air inside.

Dawson was effectively providing a plausible answer to the centuries' old story of frogs and toads in rock. It was a neat answer with proof in the form of the two halves of the split flint plus a dehydrated toad (a *Bufo* species) inside one half. Perhaps it was just a little too neat, and I have misgivings simply because of the nature of Dawson's reputation. He seems to have had a penchant for solving riddles, by coming up with the right solution at the right time. He was clearly associated with the discovery of the skull of Piltdown Man at Lewes in Sussex: a skull exhibited for thirty years in the British Museum until it was

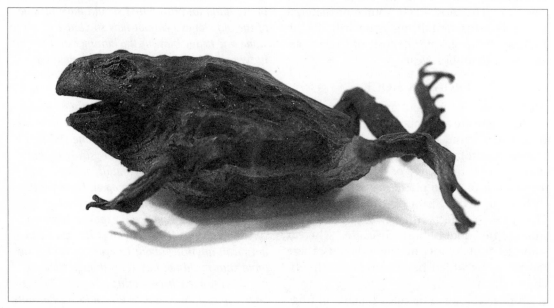

FIGURE 5 *From Forcett, Tasmania: the mummified remains of a frog (the modern species* Limnodynastes dumerilii) *found in a cavity in solid mudstone. Specimen in the collection of the Tasmanian Museum and Art Gallery.*

discovered that it was a fake, and one of the most elaborate and successful hoaxes perpetrated upon the scientific community. There was a need to find a logical explanation for an inquisitive and sceptical audience; Dawson provided it.

The second dead specimen is lodged in the collection of the Tasmanian Museum and Art Gallery at Hobart. It is a mummified *Limnodynastes dumerilii*, a burrowing frog which lives in Tasmania today (Figure 5). Accompanying the specimen is a note written by Michael Sharland of Hobart and dated 5 October 1962. The note is headed 'mummified frog' and it reads as follows:

This frog was found in a cavity in a rock in the face of a cliff at Forcett, Tasmania, in 1947. The cliff was being quarried for road metal; it was mudstone. After piecing together as much as possible of the fragments of the rock which fell from some height, the cavity was found to be the same shape and size of the frog. It looked as if at some time the frog had buried itself in mud and this had subsequently solidified. And although the frog presumably was dead its skin was still damp and its flesh soft to the touch. On exposure to the air for about two weeks the frog also solidified [mummified] and has remained in its present form since then.

There are a variety of processes that can produce fossilisation of intact animals. They include deep-freezing whereby intact animals can be trapped in glaciers or other areas of ice, mummification resulting from very rapid dehydration in an arid environment, and other processes which rapidly provide a protective medium and inhibition of bacterial putrefaction. It is therefore easy to accept a mummified specimen as a fossil if it is old enough. But it is relevant to note that the Forcett frog was moist with (apparently) a normal flesh condition. If it was dead when found, as the report indicates, it is difficult to understand why it did not rot away but remained in such good condition.

But the report by Mr Cash is clearly one of finding a live toad in a rock. And there are several other accounts brought to my attention by listeners of the Science Show, each of which has to be explained against the physio-

logical knowledge that frogs require moisture, oxygen and nourishment to remain alive.

Mr Prideaux of Corinda in Queensland wrote:

In 1931, I was mining for gold on the Etheridge, 70 miles from the railhead at Forsayth. My partner and I were working on a quartz reef about four feet in width and at a depth of 60 feet in a mine we called the 'Fingerprint' after a series of aboriginal hand prints etc. in red ochre situated in a nearby sandstone cave.

One day after we had fixed our shots in the quartz I went down and began breaking up the largest of the boulders blown out by the shots, to enable these to fit into the bucket and be raised to the surface. On cracking one such boulder I was amazed to see within a cavity of about two inches in diameter a small frog, perfectly white, and alive. The frog died soon after exposure to the air. The mine was on a hill top, but I assume that the spawn must have been carried down into the reef (how long before it would be impossible to tell) by rain water in the minute crevices of the quartz.

Yet another of the many folk who wrote with a fascinating account was Mr M.A. Munro of Namanda at Launching Place in Victoria. He wrote:

In 1937, two of us were digging a hole in which to cast a large septic tank on a property between Campbelltown and Camden in New South Wales. A large block of sandstone had to be broken up to make it small enough to lift the five feet or so to the top of the hole. In it, about 12 to 15 inches from the outside in any direction was a circular chamber about the size of a golf ball, enclosed in which was a small, grey frog toad about one-and-a-half inches in length. It did not move at first and sat groggily for about 15 minutes before moving off.

There is no confirmation of this because my partner is now dead. In the sandstone there must have been a fracture line running through the hole along which some moisture could have penetrated, but there was no way the frog could have entered nor escaped.

Although I was not as aware of zoology or geology as I now am, the experience was one which made an impact that has lasted 46 years.

Most of the frogs in rock accounts fall into two categories: those found in mines and those in large blocks of rock near the surface. Frogs certainly do occur in mines. My brother-in-law Neil Edwards is a surveyor with a special interest in surveying mines. It is a hazardous occupation but he always finds time to pick up any frogs around. In 1975 he undertook a survey of the abandoned talc mine at Gumeracha near Adelaide. The mine consisted of a series of shafts and passages at different levels but with only a single entrance —a shaft about 1 m in diameter.

He found frogs throughout the labyrinth of passages. They are listed, together with the depths of the passages in which they were found, in Table 2. Obviously a lot of frogs fall down mine shafts and survive in the damp environment there, but this does not explain how they could end up entombed within the rock itself.

How frequently are frogs found in rock? Most of the reports brought to my notice occurred 50 years ago. This does not mean that people were more gullible than they are today but, rather, it may reflect the shift from manual to mechanical methods of excavation. The only instance of several frogs being found in rocks at one site, that I have heard about, was in Tasmania. Mr P.J. Young of Clarence Park, South Australia wrote:

My father often tells the story of when he was a young man building the Tarraleah hydro-electric power station, during the 1920s or '30s and says that many frogs were found in rocks underground. When a suitable rock was found and then smashed open, out would pop a frog.

I was fascinated, particularly about what constituted a 'suitable rock', so I telephoned Mr Young senior. He explained that he and some other men were digging the foundation of the power station. They took it in turns to work at the bottom of the pit. Amongst the small boulders excavated there were occasional ones that were almost perfectly round. These were thrown up on to the surface and, said Mr Young, 'We would split them open with a knapping hammer. Not all of them contained frogs', he told me, 'but quite a few did'.

From a scientific viewpoint the existence of a living frog within solid rock is utter nonsense. Frogs have a limited life span and, like any form of life, require food, moisture and air to survive. Particularly ludicrous are reports of frogs in solid marine limestone because the rock was laid down in the sea. Some of these tales perhaps should be taken with a pinch of salt. But there remain accounts that cannot be refuted with such ease. Each of the letters I have quoted here was not solicited, but volunteered in response to a radio broadcast. Each has a ring of sincerity and authenticity. Three of the writers addressed the issue of the need for water for survival and so suggested that there may have been fractures in the rock through which it could have percolated.

Why bother to split open the rocks as Mr Young and Mr Cash did? I asked each of them. Mr Young said it was because it was well known that frogs might be inside. Mr Cash said 'I don't know. I was not bad tempered or enraged about the stone. It was there and I was 17 and had a hammer in my hand'.

TABLE TWO: Frogs found in the Gumeracha talc mine, Gumeracha, South Australia

LEVEL (FEET)	CRINIA SIGNIFERA	LIMNODYNASTES DUMERILII	LIMNODYNASTES TASMANIENSIS	LITORIA EWINGII
50	—	1	1	2
60	1	—	3	1
70	—	1	—	3
160	—	3	—	3
200	—	1	1	—

CHAPTER TWO

AN ENVIRONMENT FOR FROGS

The various species of frogs living in Australia are not distributed evenly throughout the continent. Instead there are some areas where there is considerable diversity and others where there are few species. A number of factors contribute to this situation, but by far the most significant physical factor is the reliability and amount of rainfall. The ability to withstand periods of drought is an extremely significant attribute of frogs, in a continent in which drought is a common feature.

It is customary to compare the distribution of frog species against annual mean isohyets, but probably of greater relevance are the rainfall data that take into account the inherent variability at different areas of the continent. In Figure 6A–B, 90 per cent and 10 per cent percentiles are compared. These are the figures not exceeded in 90 per cent of years for which there are data, and those reached in all but 10 per cent of years respectively. These data contribute to the variability index in Figure 6C which shows that the

PLATE 2 *Rock pool upon the escarpment at Desmond's Passage, Victoria Highway, NT. Habitat of several species of rock-dwelling frogs. (M.J. Tyler)*

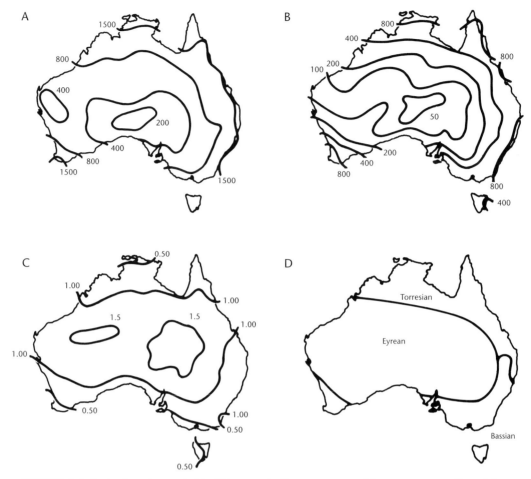

FIGURE 6 *A. Isohyets not exceeded in 90 per cent of years, expressed in mm per year. From Linacre and Hobbs (1977); B. Isohyets reached in all but 10 per cent of years, expressed in mm per year. From Linacre and Hobbs (1977); C. Annual rainfall variability expressed as a variability index. Note that the greatest variability occurs in the central, arid areas. From Linacre and Hobbs (1977); D. Biogeographic provinces and their boundaries, after Tyler (1980)*

central areas that are most arid also have the most variable rainfall. This feature, rather than actual rainfall, makes them the least hospitable environment for frogs and many other moist-dependent forms of life.

The areas of Australia with high numbers of frog species are shown in Figure 7. Some of these areas correspond to the biogeographic 'Provinces' proposed by various biologists (Figure 6D). Recognition of an area as a Province implies that it contains a unique element of species, and that there has been a period of isolation from surrounding areas sufficient for the evolution of that component.

One of the most striking features of the Australian frog fauna is the lack of dependence upon permanent bodies of fresh water. In fact, with the exclusion of a few genera exhibiting direct development and laying eggs out of water, and the specialised fauna of the eastern rainforests, the trend amongst Australian species is to breed in ephemeral pools. Thus in judging any environment in terms of its suitability for frogs, two major considerations are the soil type (whether or not it is pervious to water), and the presence within the same area of refuges in which frogs can pass the day or survive varying periods of drought. For many species the two factors are linked because the frogs involved burrow into the soil.

It is also possible to examine the various kinds of bodies of water in Australia in terms of their value as breeding sites.

BREEDING SITES

Rivers

The only feature that all rivers have in common is that they carry flowing water. Thus some may be little more than boulder-strewn areas 2 or 3 m across in rocky cuttings, whilst others are several hundred metres wide and pass through alluvial soils. But both forms are significant as frog habitats. For example, the River Murray in South Australia is the habitat of three species not occurring elsewhere within the state: *Litoria peronii* which shelters beneath the bark of flanking Eucalyptus, and *Limnodynastes fletcheri* and *Crinia parinsignifera* which live among dense aquatic vegetation at the edge of the water.

In the Kimberley Division of Western Australia, and in the Northern Territory the smaller rocky rivers arise far inland. They pass through seasonally arid country and in doing so provide a moist refuge for a number of species. Wet dependent species such as *Litoria bicolor* extend further inland than would

otherwise be the case if the rivers did not exist. Some species are confined to these rocky creek beds—for example *Litoria coplandi*, *L. meiriana* and *Megistolotis lignarius* (Plate 2).

Waterholes

The description 'waterhole' is usually applied to a depression in the bed of a temporary river. Thus, when the water level drops and the river ceases to flow a series of waterholes remain. Some Australian waterholes are vast: 200 m wide and up to 5 km long. Few frogs are to be found in waterholes, probably because these are homes to turtles and fish which feed on frogs. Nevertheless a number are found in the vegetation at the vicinity during the dry season.

Waterholes of central Australia, for example those of the Cooper Creek in South Australia, have an unstable fauna and what takes place there is a series of extinctions. Thus the river fills with a flow from Queensland and animals flood in with the water. Then, with drought the river evaporates, and eventually the waterholes dry up and remain dry until there is another cycle.

Billabongs

In the Northern Territory billabongs upon the floodplain are the equivalent of waterholes. A number of species is associated with vegetation at their periphery, but the presence of turtles and large numbers of fish within the billabongs minimises their value as breeding sites for frogs (Plate 3).

Freshwater Lakes and Ponds

There is considerable variation in the size and nature of freshwater lakes and ponds. They tend to be concentrated in the south-west of the continent and along the eastern seaboard. Because they are so highly variable in their characteristics it is difficult to generalise about their significance as breeding sites for frogs. In rainforests in Queensland as many as 15 species may breed in a single small pond. However the number and diversity of species using a single resource is not directly related to the size of the water body, whilst the survival of the offspring will be influenced by the presence or absence of fish in the water.

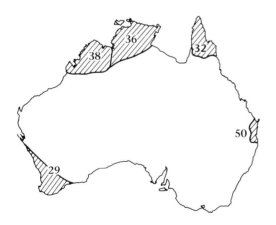

FIGURE 7 *Areas of Australia with high numbers of frog species. The boundaries of the units on the Cape York Peninsula, and in south-east Queensland are those used by Stanton and Morgan (1977).*

PLATE 3 *East Alligator River floodplain.* Litoria dahlii *was basking upon the waterlily leaves. (M.J. Tyler)*

PLATE 4 *Cracking clay at Cannon Hill, NT. In central and northern Australia numerous species seek refuge down the cracks. (M.J. Tyler)*

PLATE 5 *Claypan at Roxby Downs, SA. Habitat of* Neobatrachus centralis. *(M.J. Tyler)*

PLATE 6 *Margaret Davies and Graeme Watson dragging a net through a large temporary pool at Newry Station, NT. (M.J. Tyler)*

Gravel Scrapes

Otherwise known as 'borrow pits', these are shallow quarries excavated close to roads. They are sites where gravel and rocks have been excavated for road foundations. Because of their impervious floors the gravel scrapes fill with water and, in the absence of fishes, they provide a wonderful breeding site for frogs.

It is worth reflecting that the activity of quarrying is a positive step which has the effect of increasing the number of frogs. Too frequently conservationists view all alterations to the landscape as retrograde steps. But from time to time elements of the fauna benefit in unexpected ways.

An unfortunate habit of cattle drinking in pools is their tendency to urinate and defecate in the same water. The result is a fairly rich brew. Despite these undesirable qualities frogs seem perfectly willing to deposit their spawn there.

Farm Dams

Farm dams are most commonly occupied by frogs when there is litter at the edge. Where dams are used by cattle the surrounding ve-

getation tends to be eaten and the soil at the edges compacted. A further undesirable attribute of many dams is the presence of introduced fish.

Rock Pools

Pools upon rock faces or escarpments provide an unusual breeding site for many species. Unusual in that, while the water in the pools is normally static, during rainfall it resembles a depression in a fast flowing stream. The species that breed in these pools have tadpoles with special sucker mouths (see Chapter 7) enabling them to attach themselves to the rock floor, and so avoid being swept away.

Clay Pans

In arid areas with predominantly clay soils, water tends to lie for long periods in large, very shallow depressions. Several species of *Neobatrachus* and *Cyclorana* burrow into the damp soil at the edges of the pans and emerge at night (Plate 5).

Flooded Grassland

Because permanent areas of fresh water are so scarce in Australia, temporarily flooded areas are extremely significant breeding sites. They seem to be particularly important in northern Australia where the breeding season is dependent upon the onset of heavy rains to produce local flooding (Plate 6).

REFUGES

Another significant environmental factor influencing the distribution of frogs is the presence of adequate refuges where the frogs can hide during the day and, particularly, throughout periods of drought. In the case of the desert burrowers that bury themselves in the soil, all that matters is the presence of a suitable soil type into which they can dig. But many species hide away using existing refuges. For example, some clay soils crack as they dry, leading to their common name of 'cracking clays' (Plate 4), and several species of frog hide away in the moist layer at the foot of the cracks. At nightfall they emerge, scaling the sides like miniature mountaineers. Significant areas of cracking clays occur on the floodplains of the Northern Territory and upon the blacksoil plains of several States.

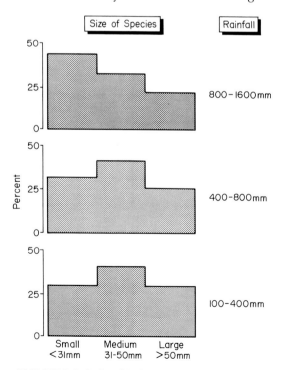

FIGURE 8 *Relationship between size of frog species and rainfall in the Northern Territory.*

Frogs have the knack of finding any crevice where the temperature is cooler than that outside. This habit was demonstrated to me at a roadhouse at Daly Waters in the Northern Territory, where I found a green tree frog, *Litoria caerulea*. It had crawled through a narrow gap between two sheets of metal. Unfortunately, the sheets were a part of an ice-making machine, and the result was one deep-frozen frog.

Cavities as different as those excavated in the soil by freshwater crabs, knot holes in timber or holes in the branches of coolibah trees—all are inhabited by frogs but, somewhat unintentionally, humans have added an extra dimension.

In northern Australia the most significant human-made frog niche is the public toilet. It provides a permanent supply of water and, at the edge of the pan, a cool refuge to hide in. Almost all *Litoria splendida* have been found in, on, or beside toilets.

In the south of the continent the greatest contribution made by humans to frogs is the habit of discarding sheets of corrugated iron. When dumped beside water, corrugated iron prevents evaporation from the soil beneath it and thus is an ideal year-round refuge.

The form, nature and frequency of water bodies and refuges are major factors influencing the number and diversity of species occurring throughout the continent. However studies in the Northern Territory have revealed that there is an association there between number, size and lifestyle with latitude. Thus the highest number of species occurs in the high rainfall periphery, and it is there that the greatest number of small-sized species are to be found (Figure 8).

Because there is an inverse relationship between body volume and surface area, small frogs have a proportionately larger surface area than large frogs. Thus small frogs have lower reserves when faced with dehydration, and in consequence few small species penetrate arid areas. Thus the incidence of terrestrial/aquatic, arboreal and fossorial species varies with latitude (Figure 9).

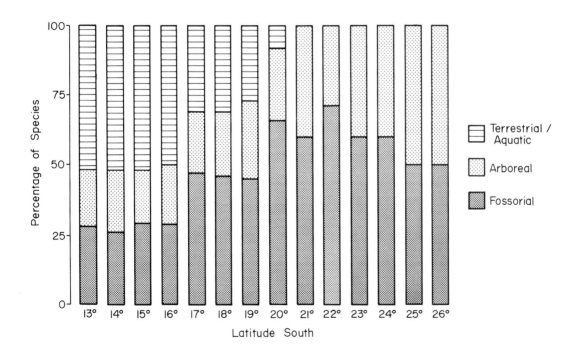

FIGURE 9 *Distribution of various life-styles of frogs in the Northern Territory in relation to latitude.*

CHAPTER THREE

THE FROG FAUNA

Australian frogs (including the introduced Bufonidae) represent five families, of which two extend throughout
the continent whereas others occupy specific
areas of the mainland. In addition to these
'free roaming' frogs a representative of a sixth
family (the clawed frog, *Xenopus laevis* of the
Pipidae from southern Africa) is maintained
in laboratories in several states. It does not
appear to have been accidentally or deliberately released into the wild.

FAMILY BUFONIDAE

The sole representative of this family in
Australia is the cane toad *Bufo marinus* which
was introduced in 1935. It occurs naturally in
South and Central America and in the
extreme south of North America. Because of
the great interest in this species it is the subject of a separate chapter (Chapter 10).

FAMILY HYLIDAE

In a recent review of the content of the family Duellman and Trueb (1986) estimated that
it includes 630 species distributed throughout
most continents, but concentrated in South
and Central America, Australia and New
Guinea. Although the majority of species live
above the ground and are appropriately
named 'tree frogs', some representatives live
on the surface of the ground, and others
beneath it.

In Australia there are three genera:
Cyclorana, Litoria and *Nyctimystes,* occurring
collectively over most of the continent.

The *Cyclorana* species have only recently
been referred to the Hylidae and are popularly called 'water-holding frogs'. They burrow
into the soil and have a habit of retaining

large quantities of water in their bodies,
which they can reabsorb during periods of
drought (see Chapter 8).

Cyclorana platycephala has a stout body
but a flattened head with small eyes, whilst
the feet are broad and have extensive webbing between the toes—an unusual feature in
a burrowing frog. In contrast the remaining
12 species share a different body form. In
them the body is roughly ovoid and the head
is triangular and high; there is little, if any,
toe webbing and the species differ from one
another in size (27–105 mm when adult) and
coloration.

In the largest species of *Cyclorana, C. australis* and *C. novaehollandiae,* most of the
bones of the skull have an external elaboration arising by the process of exostosis, that
resembles a waffle (Figure 10). Similar structural modifications characterise the skull of

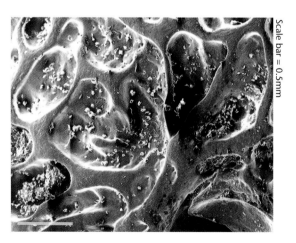

Scale bar = 0.5mm

FIGURE 10 *Elaboration of one of the bones at the
surface of the skull of* Cyclorana australis *by a
process termed exostosis (Scanning electron micrograph).*

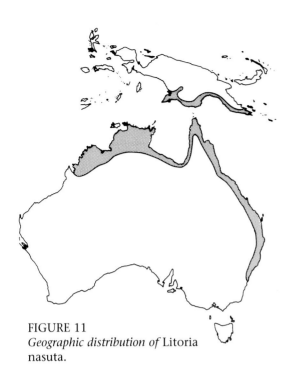

FIGURE 11
Geographic distribution of Litoria nasuta.

the crocodile, where it is believed to increase strength without the weight increase that would accompany the development of extra solid bone. The functional significance of the process in frogs is uncertain.

Although *Cyclorana* species extend as far as the coastline of northern Australia, they are absent from New Guinea, and in the south do not reach the south-west or south-east, including Tasmania.

There are over 100 named species of *Litoria* now recognised, literally dozens of new species await description, whilst no doubt many more have yet to be discovered. A final total of 150 is within the realms of possibility. *Litoria* occurs both in Australia and New Guinea (where most of the new species are to be found) and a few species are shared by both landmasses. Some, such as *L. nasuta* are distributed widely across northern and eastern Australia and have a limited distribution in New Guinea (Figure 11). Others are distributed widely throughout New

PLATE 7 *The burrowing Frog* Cyclorana australis. *Threatened, it inflates and raises its body to appear larger than it really is. (B. Stankovich- Janusch)*

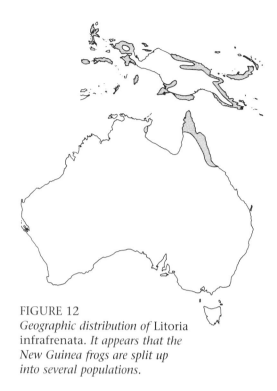

FIGURE 12
Geographic distribution of Litoria
infrafrenata. *It appears that the
New Guinea frogs are split up
into several populations.*

Several species occur upon escarpments and large rock outcrops. Most of them are active at dusk when the rock surface in summer may still be hot. One—the small *L. meiriana*—is diurnal, and a remarkable creature because of its capacity to travel rapidly in a series of bounces across the surface of water in pools upon the rock face. It would seem that the strange locomotion of the species is analogous to the 'skimming' that can be achieved by flat stones thrown at an extremely acute angle to the surface of an expanse of water. Because of their high velocity, low weight, large surface area and low angle of approach, they do not break the surface and instead bounce away from it.

But, irrespective of the explanation of the passage of *L. meiriana* across water, the marvel of witnessing this extraordinary ability for the first time, remains a memorable experience.

Around marshes and amongst the dense vegetation at the edge of rivers, streams and dams of southern Australia lives another

Guinea and adjacent islands, but have only a limited distribution in Australia such as *L. infrafrenata* (Figure 12). Thus *L. nasuta* is considered an Australian species that crossed to New Guinea, and *L. infrafrenata* a New Guinea species which invaded Australia.

Outside Australia and New Guinea *Litoria* is represented continuously as far as the Solomon Islands in the south-east and Timor in the west.

There are approximately 50 species of *Litoria* in Australia. They exhibit wide variation of body form, size, colour, habitat and biology, and several groups of species can be recognised. Whether some of these groups are so different that they merit separation as distinct genera is a matter of debate, and is discussed in Chapter 4.

Amongst the most attractive species are those that are green in life. All live in foliage, ranging from grasses and reeds in the case of *L. bicolor* to the high canopy of closed forests in the case of *L. chloris* and *L. xanthomera*. It is a green tree frog *(L. infrafrenata)* which is the largest native species in Australia and, at 135–140 mm length, it may be the largest hylid frog in the world.

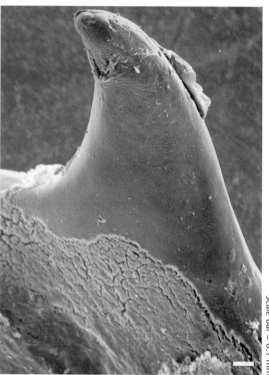

Scale Bar = 0.1 mm

FIGURE 13 *Scanning electron micrograph of one of the pair of bony tusks at the front of the lower jaw of* Adelotus brevis.

28

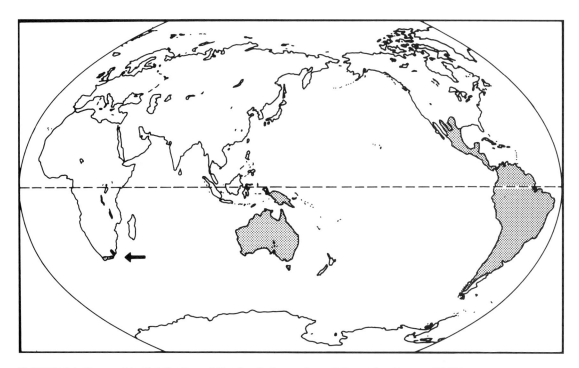

FIGURE 14 *Geographic distribution of the family* Leptodactylidae, *after Savage (1973).*

distinctive group of *Litoria* species known as the *L. aurea* group. Unlike other frogs their body surface can become slippery as the result of copious amounts of slimy and acrid secretions being liberated rapidly from skin glands. These frogs are all large, aggressive carnivores that will eat smaller individuals of their own kind as readily as other prey. The solitary northern Australian representative of this group is *L. dahlii* which tends to spend more of its time in water.

Several 'tree frogs' have reverted to a terrestrial habitat. They do not look like tree frogs at all with their elongated bodies and very long back legs. At night they move out into the open to capture insect prey.

Nyctimystes are tree frogs with two fundamental differences from *Litoria*. First, the pupil of the eye, when constricted in bright illumination forms a vertical slit in *Nyctimystes* compared with a horizontal slit or a diamond-shaped aperture in *Litoria*. Secondly, the lower eyelid of *Nyctimystes* is marked with a pattern of lines, veins or dots (a palpebral venation), whilst the lower eyelid of *Litoria* is completely transparent.

Nyctimystes centres upon New Guinea,

and its colonisation of Australia is restricted to the Cape York Peninsula of Queensland, where it was first discovered in 1967. Variability in form and colour pattern led to the belief that three species were present. However Czechura *et al.* (1987) have demonstrated that there is a single Australian species and they call it *N. dayi*.

FAMILY LEPTODACTYLIDAE

The Leptodactylidae is a family confined to the southern hemisphere, in fact largely South America, Australia and New Guinea, with a single genus *(Heleophryne)* occurring on the southernmost tip of South Africa (Figure 14).

In recent years some contributors have favoured placing the Australian and New Guinea species in a separate family: the Myobatrachidae. This matter is discussed in Chapter 4.

Australian leptodactylids are divided amongst 20 genera, of which seven each include only one species (monotypic). The first of these is *Adelotus brevis*, popularly known as the tusked frog because of the pair of bony tusks near the tip of the lower jaw

PLATE 8 *One of the hip pockets of the marsupial frog* Assa darlingtoni. *Tadpoles enter the pouch through this aperture.* *(M. Mahony)*

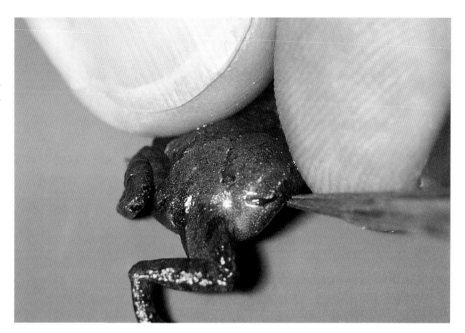

PLATE 9 Crinia tasmaniensis *of Tasmania. (B. Stankovich-Janusch)*

PLATE 10 Kyarranus kundagungan *from south-east Queensland. (Queensland Museum)*

(Figure 13). Another feature unique to *A. brevis* is the large size of the head in relation to the body. This is more marked in males than in females.

The sandhill frog *Arenophryne rotunda* is confined to a narrow zone along the middle portion of the coast of Western Australia. It was discovered in 1970 when some specimens fell into pitfall traps laid to capture small mammals for study. It is such an unusual animal that it is the subject of a separate chapter (Chapter 11).

Assa darlingtoni lives in the cool rainforests that occur around the border of southeast Queensland and north-east New South Wales. It, too, is a remarkable animal with a most unusual life history in which developing tadpoles are carried by the male in hip pockets. Had it been discovered by Charles Darwin when he visited Australia, it would now be reported in biological text books throughout the world.

The species was described as *Crinia darlingtoni* in 1933 by Arthur Loveridge of the Museum of Comparative Zoology at Harvard University. In 1966, Straughan and Main discovered that males collected in late summer had paired hip pouches in which they were carrying tadpoles and metamorphosing juveniles. Tyler (1972) erected the genus *Assa* for the species.

Ingram, Anstis and Corben (1975) were the first to report on the behaviour of the male, and the manner in which tadpoles enter the hip pouches. They observed that the eggs were laid on land and the male remained near to the clump. At the time of hatching the male became covered with egg jelly and the tadpoles swam to him upon a jelly layer and wriggled up the flanks and into one or other of the pouches (Plate 8).

More recently Ehmann and Swan (1985) provided further details of the development of the species. They noted that whereas eight to 18 eggs are laid in a clump, only five to nine found their way into the sacs of the frogs they observed. Hatching and entry into the sacs occurred after 11 days and the baby froglets emerged 48–69 days later.

The small frogs best known as froglets

31

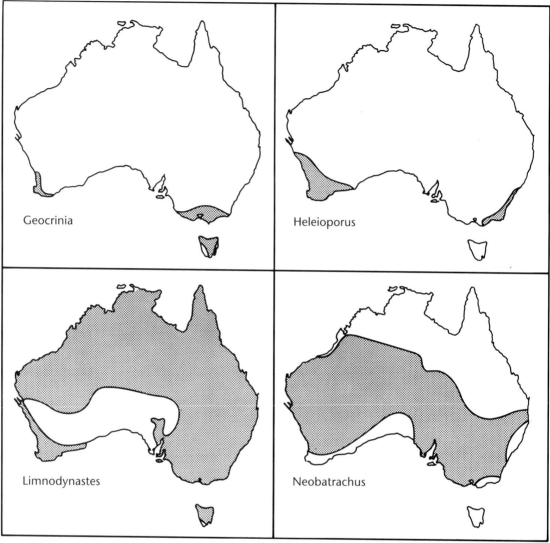

FIGURE 15 Geographic distribution of four genera of Australian frogs. Derived from data provided by Brooks (1983) supplemented by personal observations.

have had a chequered taxonomic history. From time to time they have all been lumped together in *Crinia,* first erected for the species *georgiana* of the south-west of Western Australia. All of the remaining members of the genus are substantially more slender, and have been associated in the appropriately named genus *Ranidella* ('*Rana*' = frog; '*ella*' = small one). Here I follow the latest view and lump the groups together as *Crinia* (Plate 9).

The largest number of *Crinia* species occurs in south-east and south-west Australia, but a couple of species are confined to the wet/dry tropics of the north, one *(C. desertic-*

ola) ranges widely throughout arid country whilst one *(C. remota)* extends as far as southern New Guinea.

Most *Crinia* species live in permanently damp sites or else close to water. Even the arid dwelling *C. deserticola* is most commonly found near the so-called turkey-nest dams.

Geocrinia is an assemblage of small, squat frogs that, like several other genera and groups of species, has a disjunct distribution, with some members in the south-west of Australia, and others in the south-east (Figure 15). They lay their eggs on land upon or beneath moist material and, after completing their early development, the tadpoles either remain in a sort of suspended animation, pending the onset of rain to release them from their egg capsules and wash them into pools, and then continue to develop in the normal way, or remain in the jelly mass.

Heleioporus species are moderate to large burrowing frogs with bulky bodies and short legs. Males have wicked-looking spines on their thumbs to gain a hold on the female of their choice by virtually impaling her.

Much like the distribution of *Geocrinia,* that of *Heleioporus* includes both south-east and south-west Australia but, whereas *Geocrinia* has its most extensive distribution in the south-east, *Heleioporus* has only one representative in the south-east (with a very limited distribution), and the diversity and geographic range are far more extensive in the south-west (Figure 15).

Kyarranus comprises three species of small and, in the case of one *(K. kundagungan),* spectacularly coloured frogs confined to high country upon the Great Dividing Range at the border of south-east Queensland and north-east New South Wales (Plate 10). They live mainly in cold, wet forest. Perhaps the best known is *K. sphagnicolus* of northern New South Wales. The genus is most closely related to *Philoria* of Victoria, and some contributors refer them to that genus.

Whereas 19 of the 20 genera of leptodactylid frogs in our geographical area are either confined to Australia or, if they have representatives in New Guinea, are predominant in Australia, *Lechriodus* is the exception. There are three species in New Guinea, compared with only one in Australia (Zweifel, 1972). Whether this anomaly implies that *Lechriodus* existed in New Guinea for a longer period than in Australia is, at this stage, unclear. An alternative explanation of the anomaly is that the rate of evolution and speciation in New Guinea was more rapid than in Australia over the same period.

The solitary Australian species *(L. fletcheri)* is common in coastal New South Wales from about 75 km north of Sydney to just across the Queensland border. There is a published record of *L. fletcheri* occurring at Ravenshoe in North Queensland (Moore, 1961), so bridging the huge geographic gap between the New South Wales/Queensland border and New Guinea.

Despite the convenience of this record, it has long been considered suspect. Subsequent visits by collectors to the area failed to locate a further specimen. Eventually McDonald and Miller (1982) demonstrated quite convincingly that the specimen's locality data were in error, and that most probably it had been collected at Ebor in New South Wales. Thus the huge geographic separation of the Australian from the New Guinean species is real, and the distribution of the genus remains unusual.

The gap has been explained by the finding of large numbers of the fossil *Lechriodus intergerivus* at Riversleigh Station (See Figure 4), whilst the finding of *Lechriodus casca* at Murgon indicates that the genus has existed in south-east Australia for a very long period.

According to Waite (1929), a literal translation of the name *Limnodynastes* is 'Lord of the Marshes'. Dynastes, or our word dynasty (a line of hereditary rulers) seems to have an acceptable interpretation in 'Lord', but 'Limno', the derivation of the science of limnology, is the study of all inland waters (not just marshes), including those that are salt as well as fresh (Williams, 1986).

Within *Limnodynastes* are at least three different lineages. One group includes species such as *L. dorsalis* of Western Australia and *L. dumerilii* of the south-east characterised by possession of squat, bulky bodies and a bulging gland upon the upper surface of the calf. The second includes *L. ornatus* and *L. spenceri* of northern and central Australia, characterised by highly ornate colour patterns. The remainder are genuine marsh-dwellers such as *L. tasmaniensis* and *L. fletcheri*. Collectively the members of this adaptable genus range across almost the entire continent (Figure 15).

Megistolotis is a fascinating genus represented by the single species *lignarius*. It inhabits rocky country in the north-west of the continent, and males call from beneath boulders, making an explosive 'top', closely resembling a piece of timber being struck. The mating pair produce a foam nest in which eggs are deposited.

The four Australian species of *Mixophyes* occupy the coastal fringe from the eastern border of Victoria with New South Wales to

Plate 11 *Northern barred frog.* Mixophyes schevilli *(M. Trenerry)*

Plate 12 *The turtle frog.* Myobatrachus gouldii *of WA. (M. Davies)*

Plate 13 *The undersurface of* Crinia tasmaniensis *of Tasmania (see Plate 9). (B. Stankovich-Janusch)*

Plate 14 Uperoleia aspera *exhibiting distended skin glands. (B. Stankovich-Janusch)*

the Cape York Peninsula of northern Queensland. Recently an additional species has been discovered in the Southern Highlands of New Guinea (Donellan, Mahony and Davies, 1990). It is popular these days to describe anything unusual as being enigmatic. The distribution of *Mixophyes* could be viewed as enigmatic for, whereas most genera shared by Australia and New Guinea have representatives at the north-east tip of Australia and in the southern lowlands facing Australia, *Mixophyes* of New Guinea is separated from the most northerly Australian species *(M. schevilli)* by a distance of 800 km (Plate 11).

The turtle frog *Myobatrachus gouldii* of the south-west of Western Australia is one of only two Australian species that burrow head first (Chapter 11). Its appearance gives rise to its common name, the turtle frog, being just like a baby turtle. Some of the details of its breeding biology beneath the surface of the ground have been resolved by Roberts (1981) (Plate 12).

Neobatrachus includes 10 species and has a most unusual distribution: the southern part of the mainland, extending northwards into the wet/dry tropics but poorly represented there (just reaching Derby at the southern end of the Kimberley) and absent from Arnhem Land (Figure 15). At the south it may be limited by its tolerance to winter cold, but its northern distribution includes areas exposed to intense heat, and so intolerance to heat is unlikely to be a factor explaining its minimal penetration there.

Notaden species are so gross, with such fat bodies and tiny limbs, that even a casual glance is sufficient to enable one to conclude that they have minimal ability in the jumping department. But *Notaden* never attempt to jump; instead they run, and do so with a remarkable turn of speed, such that, on occasions, I have had to run after them.

In water they are truly remarkable. Males call whilst floating in shallow pools, making a strange 'whoop whoop-whoop . . .' But the callers are skittish and at the slightest disturbance deflate rapidly and sink like stones. Four species of *Notaden* are recognised and the genus occurs throughout northern and eastern Australia.

Philoria includes only *P. frosti* and is confined to Mount Baw Baw in Victoria. It lays eggs in a foam nest in shallow water and, although the tadpoles hatch and are free living, they continue to feed upon the yolk and do not capture food of their own (Malone, 1985). *Philoria frosti* is considered endangered, largely because of its limited distribution and the competing use of the resource as a skiing resort.

The literal translation of *Pseudophryne* is 'false toad' and the members of this genus are known popularly as 'toadlets'. Small, slender bodied and short-limbed they predominate in south-western and south-eastern Australia. One of their unique features is possession of large variegated areas of pigment upon the undersurface of the body. Some *Crinia* species have similar bold markings (Plate 13).

Despite evidence that *Pseudophryne* males are aggressive to one another during the mating season, the general disposition of these animals is a shy and timid nature.

The gastric brooding frogs of the genus *Rheobatrachus* are unique in their reproductive habits, and they are discussed in Chapter 12.

In their appearance *Taudactylus* species resemble *Crinia,* and the main diagnostic feature is an internal one: instead of ending in a spike or knob, the ends of the terminal phalangeal bones are T-shaped. All of the species are small and all are confined to the eastern coastal margin of Queensland. One of them *(T. diurnus)* once abundant in south-east Queensland, has not been seen for several years and is considered an endangered species.

Whereas in 1980 just two (questionably three) species of *Uperoleia* were known, the total had jumped to 24 by 1993. These tiny, squat creatures have highly glandular skins, and many secrete toxins when handled. They differ from one another only slightly, and identification often can only be established by analysis of the male call and examination of bone structure (Plate 14).

FAMILY MICROHYLIDAE

Two microhylid genera occur in Australia: *Cophixalus* and *Sphenophryne*. Both are more speciose and more widely distributed in New

FIGURE 16 *Recently hatched microhylid frogs from Queensland upon a one cent coin.*

Guinea.

In Australia *Cophixalus* is confined to the Cape York Peninsula of Queensland. Until recently just five species were known but Zweifel (1985) added six more. Then a further two were collected by G. Werren and M. Trenerry (G. Werren, pers. comm.) bringing the total to 13. All are small species laying eggs upon land, from which tiny froglets emerge. Those shown upon a one cent coin in Figure 16 must be amongst the smallest in the world. There are five Australian *Sphenophryne*

species and all are small frogs found on the rainforest floor (Zweifel, 1985). Four, like *Cophixalus* species, are confined to the Cape York Peninsula; the fifth, *S. adelphe*, is found only on the northern margin and adjacent offshore islands of the Northern Territory, where it lives in diverse areas including mown lawns around houses.

FAMILY RANIDAE

The Ranidae is a family that dominates the frog fauna of Europe, Asia and North America; the most widely distributed and certainly best known genus is *Rana*. In outer London, Bombay or New York, a *Rana* species will be the local (and possibly only) frog. Australia is one of the last frontiers against the invasion of *Rana* species; until recently it was considered to be confined to the Cape York Peninsula, but it has now been found on the east coast of the Northern Territory.

Two names have been applied to the Australian population of *Rana*: *R. papua* from West New Guinea and *R. daemeli* from Australia. Because many frog species occur both in Australia and New Guinea there is a problem of nomenclature. However J.I. Menzies (1987) reported that there is one species that ranges eastwards from Australia through New Guinea to New Britain and that it should be called *R. daemeli*.

CHAPTER FOUR

NOMENCLATURE AND CLASSIFICATION

HYLIDAE AND PELODRYADIDAE

The business of giving frogs names, and classifying them into groups that reflect their relationship to one another, is not the most exciting activity known to mankind. Animals need names that are sufficiently distinctive to avoid ambiguity when we talk about them, and the classification will, we hope, be one that results in the creation of natural groups.

Problems occur in trying to obtain a consensus on the status of some of these units. For example, I could regard the suite of characters exhibited by a particular species of frog to be so unusual as to merit its placement in a unique genus. And if I felt that genus so unlike any other I could place it all on its own in a special family. Someone else might hold a slightly different opinion, arguing that it was certainly unique in many respects, but perhaps best regarded as a unique subfamily in an existing family.

It is differences of opinions of this kind that create uncertainty about the classification of frogs on a worldwide basis, and they are of course very difficult to resolve. In addition to problems of interpretation often there are situations where the data are inadequate to permit an unequivocal interpretation to be made.

By far the most serious, or at least significant of these quandries is the status, and so the name, of both of the major families in

PLATE 15 Litoria cooloolensis, *a small species from south-east Queensland, commonly found clinging to grass stems. (B. Stankovich-Janusch)*

PLATE 16 Litoria alboguttata *in pond. (M. Trenerry)*

Australia: traditionally known as the Hylidae and the Leptodactylidae. Whereas I have used them in Chapter 3 in their traditional way, it is important to note that some authors refer the same frogs to the Pelodryadidae and Myobatrachidae respectively. In addition some would place one leptodactylid (myobatrachid) genus, *Rheobatrachus,* in a separate family: the Rheobatrachidae.

Both the Hylidae and Leptodactylidae have extensive distributions outside Australia, and the application of these two names for the Australian representatives implies a close relationship to the frogs on other continents. To call them by other family names indicates the possession of sufficient unique characters to demonstrate that they merit separate recognition as families. A contrary view would be to say that to classify the Australian frogs otherwise than as unique units would make the Hylidae and Leptodactylidae heterogeneous units and lose their natural nature.

To change the name of a family is a fairly major step and certainly not one to be taken lightly. However the well documented argument that one would expect as a background is lacking for both the changes to the Pelodryadidae and the Myobatrachidae.

In the case of the group named the Hylidae, this name was used exclusively for almost a century. The proposal for a change occurred at a symposium on the evolutionary biology of frogs. In the subsequent publication on the contribution there, Savage (1973) maintained that the Australian frogs were so different from the frogs that made up the rest of the family that they simply should not be associated with them at the family level. Savage did not express this opinion in so many words and, in support of this deduction he cited only a study by Tyler (1971a) of vocal sacs and their associated musculature: a work that addressed the status of genera, but not of families.

Differences of opinion expressed by various contributors to that symposium produced what Duellman (1975) described as 'an array of changes in classification that are confusing to many systematists and incomprehensible to most non-systematists'. Following that situation Duellman (1975) tried to produce a classification 'reasonably compatible with evolutionary morphology of adults and larvae'. In it he did not recognise the Pelodryadidae as distinct from the Hylidae, commenting 'recognition of the Pelodryadidae offers little additional credibility to the classification of frogs'.

Duellman's contribution may be likened to a 'peace package' in the taxonomic sphere, in which he tried to find a compromise acceptable to all parties. But, to judge from subsequent events, this objective was not achieved.

In France, Dubois (1983, 1984) used the name Pelodryadidae in two contributions on the classification and nomenclature of the Anura. I questioned his decision (in correspondence) and drew attention to a review I'd published in 1979, in which I argued that the similarities between the frogs of Australia and other continents indicated that they were members of a single family (the Hylidae). Such divergence unique to the Australian frogs was, I felt, best indicated by placing them in a separate subfamily. This step acknowledged that divergence had taken place, but did not cloud their intercontinental relationships. I was expressing a personal opinion, formed from extensive comparisons of continental groups.

Dubois (1985) responded by reversing his earlier decision and supporting subfamily status within the Hylidae.

Independently, Laurent (1979) of Argentina had produced his own classification of frogs; he followed Savage (1973), recognising the Pelodryadidae. Subsequently Laurent (1985) complimented Dubois on the latter's scheme, but selected Dubois' (1985) decision to suppress the Pelodryadidae for specific criticism.

On this occasion Laurent (1985) claimed that my 1979 conclusions hinged upon biochemical studies by Robinson and Tyler (1972) that compared levels of the catecholamines adrenaline and noradrenaline in the adrenal glands of Australian frogs.

More recently Savage (1986) has provided further explanation of several novel aspects of nomenclature introduced in his 1973 contribution. He argues that the name Pelodryadidae is appropriate if the Australian frogs are to be distinguished at the family level, but he provides no further evidence to support family recognition. Finally, Hutchinson and Maxson (1987a) concluded from immunological studies that recognition of the Pelodryadinae as a subfamily was appropriate.

As an overview this may seem a long-winded account of a fairly straightforward issue about which name to use for a family of Australian frogs. But describing the events on a blow by blow basis demonstrates the paucity of data on which decisions, of great significance to the nomenclature of frogs, may be based.

LEPTODACTYLIDAE, MYOBATRACHIDAE AND RHEOBATRACHIDAE

The name to be applied to the second major group of Australian frogs is equally controversial. These, the frogs that dominate most terrestrial and fossorial niches were the subject of a major review by Parker (1940) who associated them together in the family Leptodactylidae, and recognised two subfamilies in Australia: the Cycloraninae and the Myobatrachinae.

At the time of Parker's review less than one-half of the Australian species had been discovered. Leptodactylidae was accepted by Australian zoologists until Lynch (1971) produced a worldwide review of what he termed leptodactyloid frogs, being the leptodactylids and their assumed close relatives. In a footnote on page 204 Lynch flagged the creation of a new family for the frogs of the Australian and African continents:

Elsewhere (Lynch, ms) I have proposed uniting the Old World leptodactylid subfamilies as the family Myobatrachidae and using Leptodactylidae for the four Neotropical subfamilies.

In that paper Lynch (1973) explored various arrangements of genera in subfamilies and families and proposed that the Myobatrachidae be established for the Australian subfamilies Cycloraninae and Myobatrachinae plus the South African Heleophryninae. Savage (1973) adopted the use of the Myobatrachidae but restricted it to the Australian subfamilies, and included the South African frogs in a separate family: the Heleophrynidae.

Duellman (1975) provided diagnoses of the various families recognised in his revised scheme, and comparison of those for the Myobatrachidae and Leptodactylidae (with minor corrections) suggests flimsy morphological evidence in support of their recognition:

Myobatrachidae—Presacral vertebrae 8, first and second fused or not; maxilla normally dentate; phalangeal formula normal [or reduced]; small accessory head [tendon] of M. gluteus magnus *present [or absent]. Australia and South Africa.*

Leptodactylidae—Presacral vertebrae 8, first and second [not fused]; maxilla normally dentate; phalangeal formula normal; accessory head [tendon] of M. gluteus magnus *present. South, Middle, and Southern North America, West Indies.*

A subsequent change occurred to the content of the Cycloraninae, and this affected its name. Following a series of contributions about the anomalous (hylid-like) form of the genus *Cyclorana,* Tyler (1978) demonstrated that it really was a hylid, and so transferred *Cyclorana* from the Leptodactylidae to the Hylidae. The remaining genera in the Cycloraninae took the name of the oldest genus left in the group *(Limnodynastes)* and became known as the Limnodynastinae.

That *Cyclorana* is a hylid frog with close affinities to *Litoria,* and particularly the *L. aurea* species group is now accepted, and support for this contention has been provided by biochemical studies (Maxson *et al.*, 1982).

Australian authors have embraced the term Myobatrachidae with more vigour than use of the Pelodryadidae as a substitute for the Hylidae. Most have adopted the change with little if any comment, and the shaky foundations of the Myobatrachidae pass unquestioned. An exception to the trend of not commenting is that of Hutchinson and Maxson (1987b) who state, 'We use the adjective myobatrachid as a convenient term to refer to all Australian leptodactyloid frogs'.

More attention has been devoted to the issue of the relationships of the various groups associated together. For example, Heyer and Liem (1976) considered *Rheobatrachus* sufficiently distinctive to merit recogni-

tion as a subfamily. Farris *et al.* (1982) and Tyler (1983) concluded that *Rheobatrachus* is a group distinct from, but most closely related to, the Limnodynastine. Dubois (1985) also regards the Rheobatrachinae as a leptodactylid subfamily; Laurent (1979, 1986) considered it a family: the Rheobatrachidae; whilst Daugherty and Maxson (1982) and Duellman and Trueb (1986) regard *Rheobatrachus* as just a genus in the Myobatrachidae.

GENERA AND SPECIES

Zoological nomenclature is supposed to aim at ensuring universal use and, at the same time, to maintain stability. This must not interfere with the freedom of expression and the exercise of taxonomic judgements. To oversee nomenclature there exists an International Commission and, to provide a uniform approach, there is an international code governing principles and practice.

From time to time problems are encountered that involve conflicts in interpretation between zoologists, or the need for a decision about a nomenclatural matter within the province of the Commission. Such a matter occurred in 1987 when the Australian Society of Herpetologists petitioned the Commission to suppress the impact of two privately published works on Australian herpetology and a third on that of New Zealand. The circumstances are certainly unique in Australia and are worth relating.

Cogger, Cameron and Cogger (1983) produced a catalogue of the Amphibia and reptiles of Australia. It gave their interpretation of the status and nomenclature of the Australian herpetofauna: a task of such massive proportions that it would be surprising if a number of the taxonomic decisions were not disputed by their colleagues. Nevertheless the bulk of the work met general approval.

Within just a few months Wells and Wellington (1984) produced their own 'synopsis' of Australian reptiles, clearly in response to the publication of Cogger *et al.* increasing the number of species by the staggering total of 218 and also producing 50 extra genera. Hard on its heels came a 'classification' of Australian Amphibia and reptiles (1985) with 36 new frog species, and 28 extra

genera. In that one publication they doubled the number of frog genera recognised in the previous 200 years. But even this was small fry compared with the Wells and Wellington treatment of reptiles.

The events resembled a complete spoof and were produced, the authors declared, as 'an alternative' to the scheme of Cogger *et al.* But it wasn't a spoof. The publications met the minimal standards required in the Code of Zoological Nomenclature, and representa-

tives of the professional herpetological community expressed complete condemnation of the actions of Wells and Wellington (King and Miller, 1985; Heatwole, 1985; Gans, 1985; and Tyler, 1985b, 1985c). These were followed by an application from the Australian Society of Herpetologists (1987) seeking suppression of the relevant Wells and Wellington publications in toto.

The main intention was not to gag Wells and Wellington (who must remain free to

PLATE 17 *The corroboree toadlet,* Pseudophryne corroboree *(Queensland Museum)*

nised that 'Wells and Wellington have displayed a contempt for the Code and its arbitration provisions', and that the works of these authors involved 'poor scientific and editorial practice'. (*Bulletin of Zoological Nomenclature* 48(4) Dec. 1991, pp. 337–338).

Earlier I indicated that one of the major sources of controversy is the level of significance attributed to the divergence that frogs exhibit, and noted that differences in interpretation create instability. But this has to be tolerated as part of the process of gaining knowledge and improving classification.

An example of the sort of differences of opinion I allude to occurred in the Hylidae. Tyler and Davies (1978) examined diversity in the genus *Litoria* and noted that it was disjunct rather than continuous, so permitting the recognition of groups of species sharing common features, and distinguished from other groups by the combinations of those features. We termed them 'species groups' and stated:

It is conceivable that some of those that we propose here may ultimately merit elevation to the status of separate genera. However the implications of such steps are profound, and at present it seems desirable just to propose and define the distinguishable subgeneric components that can be recognised within Litoria.

Explaining his 1973 action of recognising *Pelodryas* as a genus distinct from *Litoria,* Savage (1986) wrote:

*I was convinced that the type-species, the well-known, large, green tree frog (*Pelodryas caeruleus*) was generically distinct from the type-species of Litoria (*L. freycineti*).*

He then proceeded to list features in Tyler and Davies' (1978) definition of the *L. caerulea* species group as the diagnosis of *Pelodryas,* and, following Tyler and Davies' interpretation, included *L. splendida* in it.

The options for the classification of these Australian frogs seem almost infinite. It is not just the features of the frogs that are confusing, but also the differing interpretations of the significance of that divergence.

express their views on any topic). Rather, it was to avoid the situation where it is obligatory for zoologists to heed what they have to say and so adopt the huge number of new names scattered throughout their publications.

Four years after the application was lodged the Commission reported that it had decided not to vote on the application because it considered it to be 'outside its remit'. This was despite the fact that it recog-

CHAPTER FIVE

FOOD AND FEEDING

STUDY METHODS

There are three ways of determining the diet of an animal: watch it feeding; kill it after it has been feeding and then examine the contents of the stomach and intestine; or examine undigested material remaining in faecal pellets. Each of the methods has advantages and disadvantages to it and, increasingly, we should be concerned about practices that kill animals just to learn a little about them.

Most studies of the diet of Australian frogs have consisted of brief reviews of stomach contents of just a few specimens—Table 3—but, as shown there, three researchers examined huge samples (Pengilley, 1971a; MacNally, 1983; and Cappo, 1986). Of these MacNally examined more than 3300 specimens of just two species.

I don't intend to give the impression that it is only from large samples that useful data can be obtained. This is certainly not the case, as is well demonstrated by Calaby's (1960) study of the desert frogs *Limnodynastes spenceri* and *Cyclorana maini*,[1] reporting huge numbers of termites and ants eaten by just a few specimens of these species.

The study of the contents of faecal pellets has been attempted rarely, generally because little is recognisable, but occasionally useful data can be obtained. An example was the investigation of Tyler, Roberts and Davies (1980) on *Arenophryne rotunda* at Shark Bay in Western Australia. At the time it was not known whether the animal was rare or abundant and so it was not possible to kill specimens just to examine stomach contents. But to aid future management of populations of

this strange frog it was important to know what it ate. So from some captive specimens we collected the faecal pellets which, in frogs, are conveniently packed in little bags of mucus. Ants predominated: one faecal pellet contained the head capsules of 51 ants, whilst another contained 34.

Just what does a frog recognise as food? How does it locate food? How does it capture food?

In the case of most species it seems that a food item has to move to be recognised as food. Freshly killed insects are simply ignored. An exception is the cane toad *Bufo marinus* which will recognise as food objects that don't move, such as canned dog meat and even boiled rice (Alexander, 1965). But for almost all other species, movement is the stimulus for feeding.

FOOD LOCATION AND CAPTURE

Food location seems to be entirely by sight. I doubt if hearing contributes very much to this process. But of food capture it is possible to say a good deal more. Typically frogs capture food on their long, sticky tongues. The tongue is attached right at the front of the lower jaws and is flicked far forward so that the top surface lands upon the food item which is then drawn into the buccal cavity. The entire process can occupy just a fraction of a second.

The larger of the frog species require a greater volume of food than the small ones. There seems to be an upper limit above which adhesion to a sticky tongue just won't work and the hands are used as well to help to stuff a large food item into the buccal cavity.

Few species are capable of capturing food

---

[1] Reported as *Cyclorana cultripes*

TABLE THREE: Dietary studies on Australian frogs

FAMILY	GENUS AND SPECIES	NUMBER EXAMINED	CONTRIBUTOR
Bufonidae	*Bufo marinus* [1]	47	Zug & Zug (1979)
Hylidae	*Cyclorana australis*	430	Cappo (1986)
	Cyclorana cultripes	9	Tyler, Davies & Martin (1983)
	Cyclorana longipes	77	Cappo (1986)
	Cyclorana maini [2]	12	Main & Calaby (1957)
	Cyclorana maini [2]	3	Calaby (1960)
	Litoria aurea	?	Humphries (1979)
	Litoria bicolor	302	Cappo (1986)
	Litoria caerulea	2	Rose (1974)
	Litoria dahlii	645	Cappo (1986)
	Litoria freycineti	1	Rose (1974)
	Litoria inermis	126	Cappo (1986)
	Litoria jervisiensis	I	Rose (1974)
	Litoria nasuta	148	Cappo (1986)
	Litoria pallida	248	Cappo (1986)
	Litoria phyllochroa	3	Rose (1974)
	Litoria raniformis	?	Humphries (1979)
	Litoria rothii	213	Cappo (1986)
	Litoria rubella	219	Cappo (1986)
	Litoria rubella	1	Main & Calaby (1957)
	Litoria tornieri	143	Cappo (1986)
	Litoria verreauxi	1	Rose (1974)
	Litoria verreauxi	10	Pengilley (1976)
	Litoria wotjulumensis	13	Cappo (1986)
Leptodactylidae	*Arenophryne rotunda*	2	Tyler, Roberts & Davies (1980); Roberts (1990)
	Crinia bilingua	76	Cappo (1986)
	Crinia georgiana	?	Main (1957)
	Crinia georgiana	?	Main (1968)
	Crinia glauerti	?	Main (1957)
	Crinia insignifera	?	Main (1957)
	Crinia parinsignifera	1516	MacNally (1983)
	Crinia pseudinsignfera	?	Main (1957)
	Crinia signifera	3	Rose (1974)
	Crinia signifera	46	Pengilley (1981)
	Crinia signifera	1856	MacNally (1983)
	Geocrinia rosea	2	Harrison (1927)
	Geocrinia rosea	?	Main (1957)
	Heleioporus albopunctatus	4	Lee (1967)
	Heleioporus australiacus	1	Littlejohn & Martin (1967)
	Heleioporus australiacus	2	Rose (1974)
	Heleioporus australiacus	3	Webb (1983)
	Heleioporus australiacus	1	Webb (1987)
	Heleioporus barycragus	20	Lee (1967)
	Heleioporus eyrei	78	Lee (1967)
	Heleioporus inornatus	11	Lee (1967)
	Limnodynastes convexiusculus	33	Cappo (1986)

[1] There are numerous accounts of food items eaten by *B. marinus*. I have selected one of the more significant ones.

[2] Reported as *Cyclorana cultripes*.

FAMILY	GENUS AND SPECIES	NUMBER EXAMINED	CONTRIBUTOR
Leptodactylidae	*Limnodynastes dumerilii* [3]	2	Rose (1974)
	Limnodynastes ornatus	1	Rose (1974)
	Limnodynastes ornatus	103	Cappo (1986)
	Limnodynastes peronii	7	Rose (1974)
	Limnodynastes spenceri	9	Main & Calaby (1957)
	Limnodynastes spenceri	6	Calaby (1960)
	Metacrinia nichollsi	2	Harrison (1927)
	Mixophyes fasciolatus	I	Wotherspoon (1980)
	Myobatrachus gouldii	26	Calaby (1956)
	Myobatrachus gouldii	?	Philipp (1958)
	Neobatrachus sp.	2	Calaby (1960)
	Notaden bennetti	3	Calaby (1960)
	Notaden melanoscaphus	108	Cappo (1986)
	Notaden nichollsi	3	Calaby (1960)
	Pseudophryne australis	2	Rose (1974)
	Pseudophryne australis	2	Webb (1983)
	Pseudophryne bibroni	1	Rose (1974)
	Pseudophryne bibroni	29	Pengilley (1971)
	Pseudophryne bibroni	1	Webb (1983)
	Pseudophryne corroboree	275	Pengilley (1971)
	Pseudophryne dendyi	54	Pengilley (1971)
	Uperoleia inundata	125	Cappo (1986)
	Uperoleia russelli	6	Main & Calaby (1957)
	Uperoleia sp.	41	Webb (1983)

[3] Reported as *Limnodynastes dorsalis*

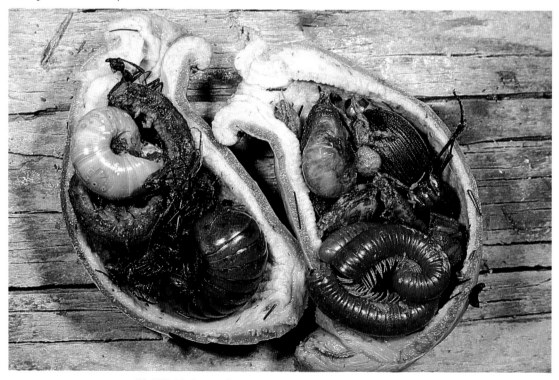

PLATE 18 *Stomach contents of* Bufo marinus. *(M. Trenerry)*

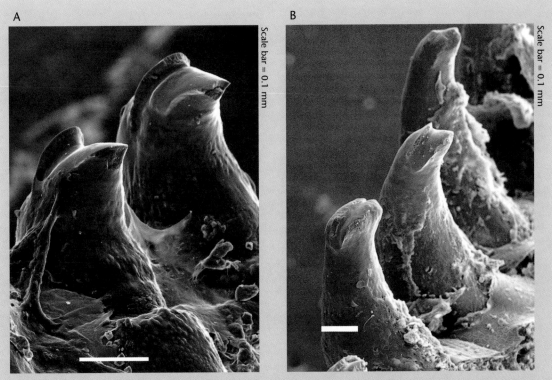

Scale bar = 0.1 mm

Scale bar = 0.1 mm

FIGURE 17 *Scanning electron microscope view of maxillary teeth:*
A. Cyclorana australis; *B.* Rheobatrachus vitellinus.

under water, but *Cyclorana platycephala* and *Litoria dahlii* can feed in this way. Both are strong muscular animals. In an aquarium they will even catch fish. The gastric brooding frog *Rheobatrachus silus* is more highly adapted to living in water, but it appears to feed upon food items at the surface or else out of the water on the adjacent soil. An adhesive tongue is of no use in water and the species there either lunge with their jaws open (rather like crocodiles) or stuff prey into the mouth with their hands. In the case of the two *Rheobatrachus* species the tongue is vestigial, fused to the floor of the mouth, and certainly unable to be flicked out of it (Horton, 1982a).

TONGUES AND TEETH

In humans tongues get scarcely any attention at all. An occasional painful scrubbing with a tooth brush (aimed at reducing the socially undesirable feature of unsightly dead cells), is the best that they can hope for. In contrast, the teeth get examined at six-monthly intervals by experts, and are scraped, polished, buffed, drilled, filled and capped. Even if they

have to be removed, they can be replaced by high-tech, artificial look-alikes. Clearly the tongue is a second-rate citizen in the human buccal cavity.

Curiously, in the frog world the relative significance of these two structures is reversed. Thus, whereas the tongue is generally the most significant organ of food capture, teeth are almost an optional extra. In fact frogs have such minimal use of teeth that scarcely any have them in the lower jaw, whilst many species get by without any at all. What's more, they survive as well in the dietary stakes as their toothed colleagues.

Species with teeth differing most conspicuously from the general form are the two *Rheobatrachus*. Those of *R. silus* were reported by Liem (1973) to be 'fang-like'. High resolution electron micrographs provided by Davies (1983) illustrate elongated and distinctly curved structures, with blunt or slightly serrated tips. In Figure 17 the teeth on the upper jaw (maxilla) of *Cyclorana australis* are compared with those of *Rheobatrachus vitellinus*.

Almost all frogs have additional 'vomerine' teeth, so named because they are located

47

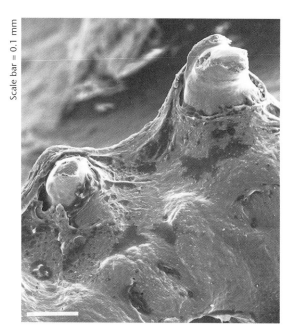

Scale bar = 0.1 mm

FIGURE 18 *Scanning electron microscope view of vomerine teeth of* Cyclorana australis.

The only real oddity in tooth structure amongst Australian frogs are the fangs of the appropriately named 'Tusked Frog', *Adelotus brevis*. These are slightly curved and sharply pointed structures (Figure 13), and when the jaws are closed they fit into special grooves in the upper jaw. The function of these fangs remains unknown. It is tempting to speculate that because the male remains with the spawn clump after its deposition, the fangs might be used to fight off marauders. But females don't remain with the spawn, yet they too have fangs (admittedly smaller ones).

Furthermore, there are few predators of spawn. Fish will certainly eat them, but other than fish and ants there are no known candidates.

Technically the fangs are termed 'odontoids' which means that they are tooth-like but not true teeth.

Whereas our teeth are differentiated for various purpose—incisors for chopping, molars for crushing and grinding—amphibian and reptile teeth are of a single kind throughout one or both jaws: what is termed 'homodont dentition'.

Frogs don't use their teeth to chop up food items into smaller portions, and there is no mastication of food whatsoever. Thus the

on the vomerine bone of the palate, near the front of the upper jaws (Figure 18). They emerge on each side from a boney protuberance, and are rather variable, even in number between the left and right sides. The teeth tend to be poorly developed or even absent in small species, and best developed in large ones.

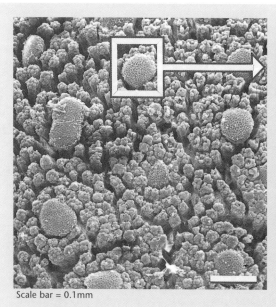

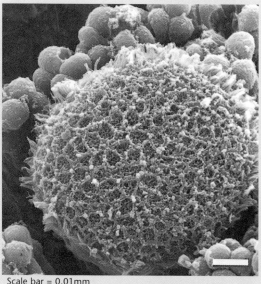

Scale bar = 0.1mm Scale bar = 0.01mm

FIGURE 19 *Scanning electron microscope view of taste buds on the upper surface of the tongue of* Litoria rothii.

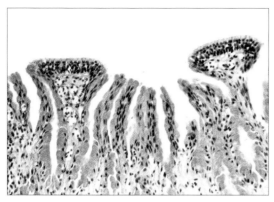

FIGURE 20 *Photomicrograph of transverse section through the tongue of* Litoria rothii. *The expanded structures are taste buds.*

only possible role of the teeth on the jaws, and vomerine bones too, is one of gripping large prey items until they can be pulled in by the tongue, or pushed in by the hands.

To the naked eye the frog tongue resembles a smooth, pink disc. However, microscopically, the surface is deeply pitted and covered with taste buds (Figures 19–20). What's more, the two muscles that are largely responsible for flicking the tongue out and withdrawing it (the genioglossus and hyoglossus), are arranged in a complex manner. They insert into the undersurface of the body of the tongue as tiny separate slips that pass between each other like interlocked fingers (Figure 21): the 'interdigitation' of Horton (1982a).

DIET

We tend to restrict the word 'diet' to the taking of a restricted range and quantity of food items to achieve weight-loss, or perhaps as a punishment, 'the prisoner was given a bread and water diet', or for some other purpose. Reference to the diet of animals has a broader interpretation: the nature of the range of food

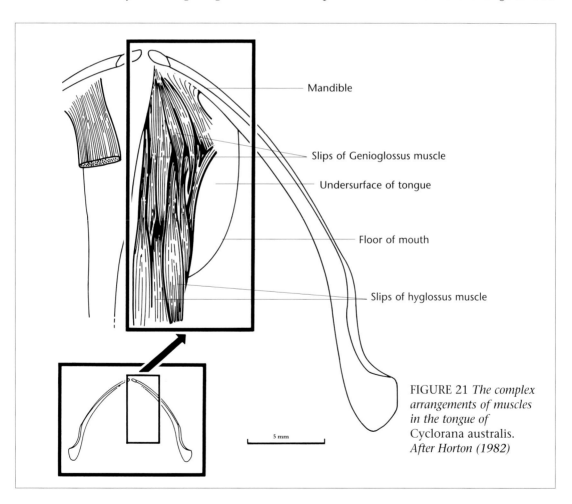

Mandible

Slips of Genioglossus muscle

Undersurface of tongue

Floor of mouth

Slips of hyglossus muscle

5 mm

FIGURE 21 *The complex arrangements of muscles in the tongue of* Cyclorana australis. *After Horton (1982)*

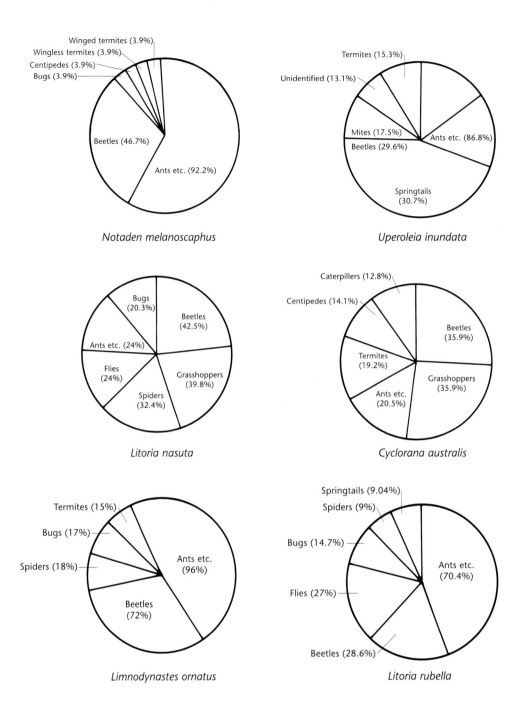

FIGURE 22 *Frequency with which major kinds of prey were found in the stomachs of six species of frogs at Jabiru, Northern Territory. From Cappo (1986)*

items ingested. It is also concerned with any limitations to that range in terms of size and diversity of food items, and whether frogs select some kinds of food items and avoid others.

Because food items are not chopped into pieces by the teeth, the upper size limit of a prey item must be the gape of the mouth. The smallest dimension of a prey item (its head or body width) must pass within the frog's jaws.

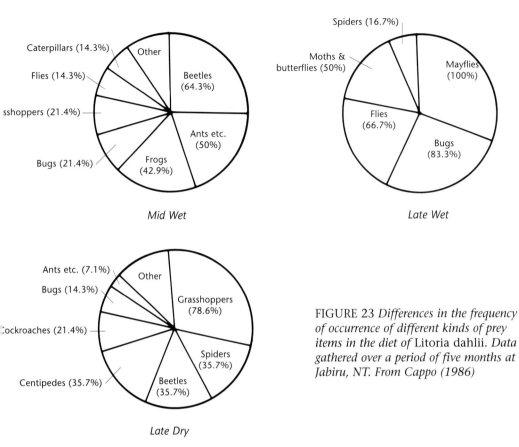

Caterpillars (14.3%)
Other
Beetles (64.3%)
Flies (14.3%)
sshoppers (21.4%)
Ants etc. (50%)
Bugs (21.4%)
Frogs (42.9%)

Mid Wet

Spiders (16.7%)
Moths & butterflies (50%)
Mayflies (100%)
Flies (66.7%)
Bugs (83.3%)

Late Wet

Ants etc. (7.1%)
Other
Bugs (14.3%)
Grasshoppers (78.6%)
Cockroaches (21.4%)
Spiders (35.7%)
Centipedes (35.7%)
Beetles (35.7%)

Late Dry

FIGURE 23 *Differences in the frequency of occurrence of different kinds of prey items in the diet of* Litoria dahlii. *Data gathered over a period of five months at Jabiru, NT. From Cappo (1986)*

Thus a marsh frog *(Limnodynastes tasmaniensis)* has been known to try to eat a baby brown snake *(Pseudonaja textilis)* six times its own body length, and a green tree frog *(Litoria caerulea)* can eat a small rat.

Assuming that the items of equal size have equal nutritional value, and energy expended in capture was equal for all items, there would be benefit to be gained from the capture of a few large items, rather than numerous small ones. In practice there is a tendency for the larger species to consume larger food items, and for small species to eat smaller ones. However in the case of the small (25–30 mm) species *Crinia signifera* and *C. parinsignifera,* MacNally (1983) found that the food items consumed most frequently were up to 1.2 mm long: that is, items that are considerably smaller than the gape of the frogs.

Numerous factors influence the range of prey consumed. By far the most significant is the habitat and season and, therefore, the variety of invertebrate prey living there. Thus, beside water frogs will encounter more species with an aquatic stage in their life-cycles, than species living further away from water. From time to time detailed knowledge of such phenomena can have a definite application. For example, at the Ranger uranium mine at Jabiru, in the Northern Territory, the construction of a huge tailings dam, and equally large retention ponds, created a new, permanent, aquatic environment. Because frogs would colonise this new resource, the question was raised whether any of these frogs would be likely to transfer radionuclides out of the area as a result of eating aquatic food items. Studies by Tyler and Cappo (1983) nominated *Litoria dahlii* as the most likely candidate, because it consumed more insects that spent part of their life-history in water.

The range of prey consumed by frogs can be expressed in the form of pie diagrams. In

Figure 22 is shown the major kinds of prey consumed by six of the species at Jabiru and the frequency with which they were found in stomachs.

Very clearly, seasonal changes in the nature and diversity of prey (availability) will limit the prey that can be eaten. For example, juicy soft-bodied termites swarm on certain warm, very humid evenings early in the wet season. On one night there may be virtually millions of them flying and covering the ground surface; on the next night there may be none at all. Stomach contents of frogs foraging in the area will vary tremendously from just one night to the next.

The predominance of particular items of prey also varies from season to season, according to the stages of development of insects and their abundance. This phenomenon is revealed in Figure 23 showing the difference in diet of samples of *Litoria dahlii* taken at intervals over a period of five months.

The way in which frogs swallow their food is poorly understood, whereas the process of ingestion is simple enough, resulting in the prey item usually being stuck to the upper surface of the tongue towards the back of the buccal cavity. Almost inevitably during the process of swallowing, the frog's eyes descend and disappear from view. The eyes are not housed in bony sockets like ours, but in fact are separated from the buccal cavity just by a thin sheet of soft tissue. Thus when they descend it is into the buccal cavity, and their hard undersurface probably comes in contact with the tongue. So for the food item to be transferred further back where it can be swallowed probably involves the tongue sliding against the eyes.

To judge from some of the tough-bodied prey ingested, such as large beetles or snails, or small vertebrates with bony skeletons, frogs must have powerful digestive processes. Studies by Taylor, Tyler and Shearman (1985) have shown that frogs do not secrete hydrochloric acid constantly in the stomachs, but only in an almost immediate response to swallowing food.

Given all of this information about food and feeding, the fundamental question to be answered is whether frogs are selective or indiscriminate feeders. The evidence of Calaby (1960) and Cappo (1986) is that from time to time some individuals will eat large numbers of particular kinds of prey. But selection means far more than simply one or more kinds of prey items. Selection implies that some prey kinds are eaten whereas others are avoided. So to be able to establish whether an animal is a selective feeder requires a simultaneous survey of the absolute availability of prey in the community. In practice this is a daunting task, but MacNally (1983) included that aspect in his research and showed that the frogs tended to eat particularly small food items, and that small items did predominate in the community. Because they appeared to be positively avoiding large food items, selection occurred, but for size rather than the nature of prey.

CHAPTER SIX

COMMUNICATION AND COURTSHIP

MEANS OF COMMUNICATION

Whereas we can communicate to one another by speech, facial expressions and by body language as well, frogs lack facial muscles and instead communicate by voice and occasionally by body language too.

Getting back to basics it is possible that frogs may well have been the first animals to communicate to one another by sound. But it seems that frog society is rather chauvinistic, for it is only the male which has a voice and, under normal circumstances, the female remains silent.

Each species has a different call repertoire, and calls can perform several functions.

KINDS OF CALLS

The call most likely to be heard is that produced in the mating season when the males congregate together and then sing their little hearts out. Each year at that time I get telephone calls from distressed folk in suburbia losing sleep because there is a chorus from a pond near their bedroom.

Each of the calls within the chorus is designed to attract female frogs and is appropriately termed the 'mating call'.

But males will call at periods of the year outside the mating season. Clearly this activity is not designed to attract females, and this call is termed the 'advertisement call'. There is a tendency these days to lump the mating call and the advertisement calls together; both are advertising the presence of the male, and the listener is not required to deduce the purpose.

The onset of rain in the daytime elicits a spontaneous response, and the reason for this response is obscure. It is termed the 'rain call'. In my garden at Belair in South Australia, switching on the lawn sprinkler results in several *Litoria ewingi* tree frogs calling from different positions in the splattered vegetation.

The so-called 'release call' is the sound emitted by a male that has been grasped by another male bent on a sexual experience. What the subject of the amorous advance is saying is, basically, that a fundamental error has been made in the choice of partner.

All of the above calls are made with the mouth tightly closed. There remains an additional call made with the mouth wide open. Perhaps appropriately this is termed the 'distress call'. Quite what purpose the distress call serves is uncertain. On the occasions when I have heard it, I've rushed outside only to find a cat sitting almost mesmerised, its nose a few centimetres from a little frog. The frog has his mouth wide open and is squealing his distress. Rather than having the effect of deterring a cat or dog, it seems to enthral and totally captivate them. They sit fascinated watching the frog closely. When the frog stops screaming they are likely to gently stick out a paw to touch it and so produce another burst of screams. The frog is not being hurt in any way but this is hard to believe when you hear the sounds that it makes.

Just whom the screams are intended to influence is uncertain. Domestic cats and dogs, who produce this response, are very recent introductions into Australia, and it is curious that the few frogs I've seen being swallowed by traditional native enemies, such as snakes, often are silent.

PLATE 19 *Male frog with inflated vocal sac. (D.&C. Frith)*

The Larynx

The organ of sound production is the larynx. In frogs the process of creating sound is quite simple and resembles the means by which you, as a child, produced sound by holding a blade of grass or a gum leaf between cupped hands, and blew through the tiny gap between the thumbs.

The larynx is shaped like an Aussie-rules football divided into two halves along its long axis. Between the two halves are the vocal cords, each of which is attached at each end to the laryngeal wall and, by a short strip of tissue halfway along their length as well.

The shape of the larynx is controlled by two pairs of muscles: one pair along its length, and the other around it. Slight shifts in the contraction and relaxation of these muscles alter the stresses upon the vocal cords and, no doubt, occur during sound production.

Vocal Sacs

Most male frogs have a special inflatable bag located under the lower jaw, termed the 'vocal sac'. This sac opens into the mouth by a pair of slit-like apertures or open holes located on each side of the tongue; during calling the sac is fully inflated and functions as an amplifier.

The role of the sac in sound amplification is demonstrated by comparing the sound produced by a species possessing a vocal sac, with that produced by a species lacking one. In the absence of a vocal sac the advertment call is soft and difficult to hear even at a distance of just 3 m. (for example, *Litoria lesueuri* reported by Moore (1961)). In contrast a chorus of six to eight *Litoria rothii* in a pool on a flat plain at Newry Station in the Northern Territory, could be heard at distances of up to 1200 m. On another occasion in open forest on Groote Eylandt a chorus of about one dozen *Uperoleia inundata* could be heard at a distance of approximately 1 km (Tyler, Davies and Watson, unpublished).

Vocal sacs differ both in shape and position (Plate 19). Most commonly there is a single lobe (unilobular) in a median position under the mouth (Figure 24A). Less commonly they have two lobes (bilobular) which inflate externally as separate pouches (as in *Rana daemeli* of north Queensland and the Northern Territory, Figure 24B). Alternatively the vocal sac may be bilobular in its structure, but inflate internally so that the entire body becomes swollen (as occurs in all *Notaden* species).

TABLE FOUR: Published analyses or detailed description of calls

GENUS	SPECIES	REFERENCE
Arenophryne	*rotunda*	Roberts, 1984
Assa	*darlingtoni*	Straughan & Main, 1966
Bufo	*marinus*	Loftus-Hills & Johnstone, 1970
Cophixalus	*bombiens*	Zweifel, 1985
	concinnus	Zweifel, 1985
	crepitans	Zweifel, 1985
	hosmeri	Zweifel, 1985
	infacetus	Zweifel, 1985
	neglectus	Zweifel, 1985
	ornatus	Zweifel & Parker, 1977; Zweifel, 1985
	saxatilis	Zweifel & Parker, 1977; Zweifel, 1985
Crinia	*bilingua*	Martin *et al.*, 1980
	deserticola	Liem & Ingram, 1977; Tyler *et al.*, 1981c
	georgiana	Ayre *et al.*, 1984
	glauerti	Littlejohn, 1959; Littlejohn, 1961
	insignifera	Littlejohn, 1957; Littlejohn, 1959; Littlejohn, 1961; Bull, 1978
	parinsignifera	Littlejohn, 1958; Littlejohn, 1959b; Littlejohn, 1961; Littlejohn & Martin, 1965; Straughan & Main, 1966; Littlejohn, 1968; Loftus-Hills & Johnstone, 1970; Loftus-Hills & Littlejohn, 1971b; Littlejohn *et al.*, 1985
	pseudinsignifera	Littlejohn, 1959b; Littlejohn, 1961; Bull, 1978
	remota	Tyler & Parker, 1974
	riparia	Littlejohn & Martin, 1965; Odendaal *et al.*, 1983, 1986
	signifera	Littlejohn, 1958; Littlejohn, 1959; Littlejohn, 1961; Littlejohn, 1964; Littlejohn & Martin, 1965; Straughan & Main, 1966; Littlejohn, 1968; Loftus-Hills & Johnstone, 1970; Littlejohn & Martin, 1974; Littlejohn *et al.*, 1985
	sloanei	Littlejohn, 1958, Littlejohn, 1959; Littlejohn, 1961; Straughan & Main, 1966; Littlejohn, 1968
	subinsignifera	Littlejohn, 1957; Littlejohn, 1959; Littlejohn, 1961
	tinnula	Straughan & Main, 1966
Cyclorana	*australis*	Tyler & Martin, 1975
	brevipes	Tyler & Martin, 1977
	cryptotis	Tyler & Martin, 1977

GENUS	SPECIES	REFERENCE
Cyclorana	*cultripes*	Tyler & Martin, 1977
	maculosus	Tyler & Martin, 1977
	maini	Tyler & Martin, 1977
	manya	Van Beurden & McDonald, 1980
	novaehollandiae	Tyler & Martin, 1975
	vagitus	Tyler *et al.*, 1983
Geocrinia	*alba*	Roberts *et al.*, 1990
	laevis	Littlejohn & Martin, 1964; Littlejohn, 1968; Littlejohn *et al.*, 1971; Littlejohn & Watson, 1973; Littlejohn & Watson, 1974; Littlejohn & Watson, 1976; Gartside *et al.*, 1979; Harrison & Littlejohn, 1985; Littlejohn & Watson, 1985
	lutea	Roberts *et al.*, 1990
	victoriana	Littlejohn & Martin, 1964; Littlejohn, 1968; Littlejohn & Martin, 1969; Loftus-Hills & Johnstone, 1970; Littlejohn *et al.*, 1971; Littlejohn & Watson, 1974; Littlejohn & Watson, 1976; Gartside *et al.*, 1979; Littlejohn & Harrison, 1981; Littlejohn, 1982; Harrison & Littlejohn, 1985; Littlejohn & Harrison, 1985; Littlejohn & Watson, 1985
	vitellina	Roberts *et al.*, 1990
Heleioporus	*albopunctatus*	Littlejohn, 1959; Lee, 1967; Bailey & Roberts,1981
	australiacus	Lee, 1967; Littlejohn & Martin, 1967
	barycragus	Littlejohn, 1959[1]; Lee, 1967; Bailey & Roberts, 1981
	eyrei	Littlejohn, 1959; Lee, 1967; Bailey & Roberts, 1981
	inornatus	Littlejohn, 1959; Lee, 1967; Bailey & Roberts, 1981
	psammophilus	Littlejohn, 1959; Lee, 1967; Bailey & Roberts, 1981
Limnodynastes	*dorsalis*	Loftus-Hills & Johnstone, 1970; Martin, 1972
	dumerilii [2]	Littlejohn & Martin, 1965
	dumerilii insularis	Littlejohn & Martin, 1974
	dumerilii interioris	Martin, 1972
	peronii	Littlejohn & Martin, 1965
	tasmaniensis	Littlejohn & Martin, 1965; Loftus-Hills & Johnstone, 1970; Loftus-Hills & Littlejohn, 1971a; Littlejohn & Roberts, 1975; Martin & Tyler, 1978; Tyler *et al.*, 1983
	terraereginae	Martin, 1972
Litoria	*aurea*	Loftus-Hills & Johnstone, 1970
	bicolor	Straughan, 1969
	chloris	Davies, McDonald & Adams, 1986
	citropa	Littlejohn *et al.*, 1972
	eucnemis	Tyler & Watson, 1986

[1] As *australiacus*
[2] As *dorsalis*

GENUS	SPECIES	REFERENCE
Litoria	*ewingi*	Littlejohn, 1965; Martin & Littlejohn, 1966; Loftus-Hill & Johnstone, 1970; Loftus-Hills & Littlejohn, 1971a; Loftus-Hills & Littlejohn, 1971b; Watson *et al.*, 1971; Littlejohn & Martin, 1974; Littlejohn, 1976; Littlejohn, 1982; Littlejohn & Watson, 1983; Watson *et al.*, 1985
	fallax [3]	Straughan, 1969
	inermis	Davies *et al.*, 1983
	jervisiensis	Martin & Littlejohn, 1966; White *et al.*, 1980
	latopalmata	Davies *et al.*, 1983
	pallida	Davies *et al.*, 1983
	peronii	Martin *et al.*, 1979
	paraewingi	Watson *et al.*, 1971; Littlejohn, 1976; Littlejohn & Watson, 1983
	pearsoniana	McDonald & Davies, 1990
	revelata	Ingram *et al.*, 1982
	rothii	Martin *et al.*, 1979
	rubella	Lindgren & Main, 1961
	tornieri	Davies *et al.*, 1983
	tyleri	Martin *et al.*, 1979
	verreauxi	Littlejohn, 1965; Loftus-Hills & Johnstone, 1970; Loftus-Hills & Littlejohn, 1971b; White *et al.*, 1980; Littlejohn, 1982; Watson *et al.*,1985
	xanthomera	Davies *et al.*, 1986
Megistolotis	*lignarius*	Tyler *et al.*, 1979
Myobatrachus	*gouldii*	Roberts, 1981
Neobatrachus	*albipes*	Roberts *et al.*, 1991
	aquilonius	Tyler *et al.*, 1981; Mahony & Roberts, 1986
	centralis	Littlejohn, 1959
	fulvus	Mahony & Roberts, 1986; Roberts & Majors, 1993
	kunapalari	Mahony & Roberts, 1986
	pelobatoides	Littlejohn, 1959; Mahony & Roberts, 1986
	pictus	Roberts, 1978
	sutor	Littlejohn, 1959
	wilsmorei	Littlejohn, 1959
Philoria	*frosti*	Littlejohn, 1963
Pseudophryne	*bibroni*	Pengilley, 1971; McDonnell *et al.*, 1978
	corroboree	Pengilley, 1971
	dendyi	Pengilley, 1971
	semimarmorata	Littlejohn & Martin, 1969; McDonnell *et al.*,1978
Rana	*daemeli*	Menzies, 1987

[3] As *glauerti*

GENUS	SPECIES	REFERENCE
Rheobatrachus	*silus*	Tyler, 1983
Sphenophryne	*adelphe*	Zweifel, 1985; Tyler *et al.*, 1991
	fryi	Zweifel, 1985
	gracilipes	Zweifel, 1985
	pluvialis	Zweifel, 1985
	robusta	Zweifel, 1985
Taudactylus	*acutirostris*	Ingram, 1980
	eungellensis	Richards & James, 1992
	liemi	Ingram, 1980
	rheophilus	Ingram, 1980
Uperoleia	*altissima*	Davies *et al.*, 1992
	arenicola	Tyler *et al.*, 1981a
	aspera	Tyler *et al.*, 1981b
	borealis	Tyler *et al.*, 1981 a
	crassa	Tyler *et al.*, 1981a, 1987
	fusca	Davies *et al.*, 1986
	glandulosa	Davies *et al.*, 1985
	inundata	Tyler *et al.*, 1981a
	littlejohni	Davies *et al.*, 1992
	laevigata	Robertson, 1984[4] Davies & Littlejohn, 1986
	lithomoda	Tyler *et al.*, 1981a, 1981c, 1987; Davies *et al.*, 1986
	martini	Davies & Littlejohn, 1986
	mimula	Davies *et al.*, 1986
	minima	Tyler *et al.*, 1981a
	mjobergi	Tyler *et al.*, 1981b
	rugosa	Davies & McDonald, 1985; Davies & Littlejohn, 1986
	talpa	Davies & Martin, 1988
	trachyderma	Tyler *et al.*, 1981c
	tyleri	Davies & Littlejohn, 1986

[4] As *rugosa*

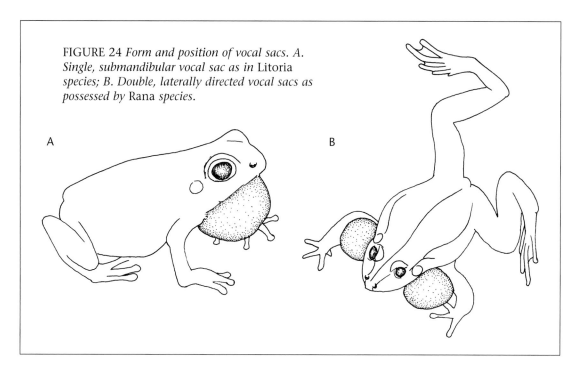

FIGURE 24 *Form and position of vocal sacs. A. Single, submandibular vocal sac as in* Litoria *species; B. Double, laterally directed vocal sacs as possessed by* Rana *species.*

All species of the leptodactylid burrowing frogs *Heleioporus* and *Neobatrachus* lack vocal sacs but, nevertheless, are able to produce quite loud calls. It seems that the buccal cavity (which is large and domed in these genera) functions as a resonance chamber, so making a vocal sac superfluous.

Several species of *Litoria*, *Nyctimystes* and *Taudactylus* also lack vocal sacs; all except *L. coplandi* occur in eastern Australia, and almost all live at the edge of rocky streams where there is a high background noise of flowing water. This background noise seems to be sufficient to mask any frog call. To attract a mate the males must use another means of communication.

Amongst Australian species the development of vocal sacs has been studied only in *Limnodynastes tasmaniensis* (Tyler, 1975) but the sequential pattern is probably similar in the other species that have a single, median sac. Vocal sac development commences with the formation of a tiny outgrowth of the floor of the buccal cavity each side of the tongue. The outgrowths enlarge progressively and become pouches. These two pouches develop towards one another and, when they meet, the dividing wall breaks down and a vast single pocket is formed (Figure 25).

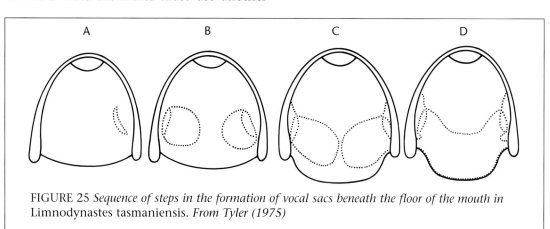

FIGURE 25 *Sequence of steps in the formation of vocal sacs beneath the floor of the mouth in* Limnodynastes tasmaniensis. *From Tyler (1975)*

59

Vocal Sac Shape

Although the shape of the inflated vocal sac is determined by the position it occupies, it has been observed in *Limnodynastes tasmaniensis* that an individual may alter the shape by constricting the muscle that forms the front part of the wall of the sac (Tyler, 1971b) as shown in Figure 26.

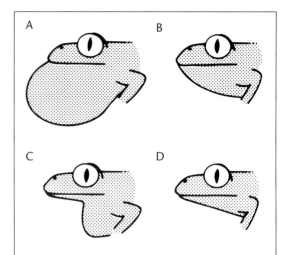

FIGURE 26 *Various shapes that the vocal sac can form as a result of voluntary control during calling by* Limnodynastes tasmaniensis. *A. Fully inflated; B. Partially deflated; C. Posteriorly inflated; D. Entirely deflated. From Tyler (1971b)*

CALLING SITES

The positions occupied by calling males vary from species to species. If several species are calling to attract females to a single breeding site it makes good sense to space them out into different areas.

Just where the male frogs call from when attempting to attract females is by no means haphazard. The spatial separation of eight species coexisting at a breeding site in the Northern Territory is shown in Figure 27. As demonstrated there, few potential calling sites are not used, and few species are competing with others for a similar site.

The following notes explain the dynamics of the species shown in Figure 27.

Uperoleia inundata calls from the base of spinifex or similar tall grasses at the edge of water. For most of the year *Litoria caerulea* calls from high positions in trees. But during the breeding season it comes down to the ground at night, and calls whilst seated upon small boulders or other slightly elevated posi-

tions. *Cyclorana australis* is such a heavyweight that it dominates the edge of the water. *Litoria bicolor* is a nimble mountaineer which calls whilst grasping grasses in or at the edge of water. *Limnodynastes ornatus* calls whilst floating at the surface of the water. *Litoria nasuta* is one of several species that call from amongst vegetation at the water's edge. *Litoria microbelos* is perhaps the tiniest of the

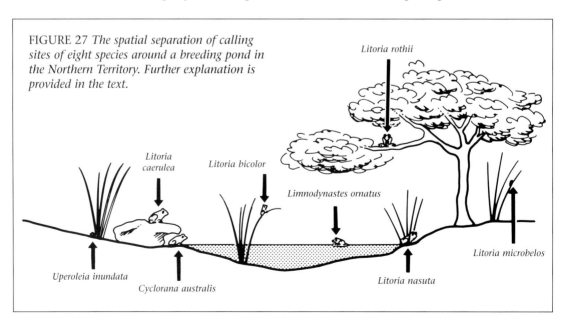

FIGURE 27 *The spatial separation of calling sites of eight species around a breeding pond in the Northern Territory. Further explanation is provided in the text.*

Litoria rothii

Litoria caerulea

Litoria bicolor

Limnodynastes ornatus

Litoria microbelos

Uperoleia inundata

Cyclorana australis

Litoria nasuta

Australian frogs. If *L. bicolor* occurs at the same site, then *L. microbelos* becomes the underdog and calls from the base of poolside grasses but if *L. bicolor* is absent, *L. microbelos* calls from the top of grass stems. *Litoria rothii* calls commonly from branches above water. In the absence of *C. australis* they may descend to the ground.

THE RESPONSE OF THE FEMALE

Before the behaviour was examined in any detail, it was assumed that males called in a chorus, females were attracted to them and, when they arrived on the scene, were grabbed by the nearest male. But in reality the inter-action is much more subtle. It seems that the female selects the male she intends to mate with. Robertson (in press) plotted the move-ments of female *Uperoleia laevigata*[1] approach-ing the mates of their choice. He summarised their activity as follows: 'Females...spend three or four nights moving slowly through the aggregations of males apparently listen-ing to a sample of males before initiating amplexus with one of them'.

Observations confirming that the female selects her mate were made at Jabiru, Northern Territory. One night there, at the height of the wet season, *Uperoleia inundata* were calling from flooded vegetation beside the Arnhem Highway. One male called from an exposed site at the base of a clump of spinifex, and he continued uninterrupted when lit by a torch beam. Thus in the light it was possible to watch a female approach him from behind. Even when she was directly behind him and separated by no more than 1 cm, the male gave no sign of being aware of her presence. The female therefore took the initiative by pushing beneath his hindlegs and emerging in front of him. Whereupon the male deflated his vocal sac and grabbed her. Such is the height of the desire of the male that he will grab at most anything. Cane toads have been observed grasping a table-tennis ball, small tree stumps, and even dead toads upon the road. Thus the occasional mis-take in which males grab the wrong species is to be expected (Plate 20).

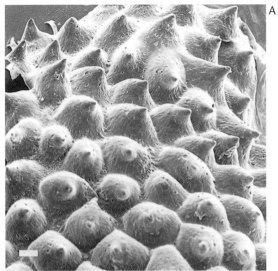

Scale bar = 0.1mm

Scale bar = 0.01mm

Scale bar = 0.01mm

FIGURE 28 *Scanning electron microscope views of the surface of the male nuptial pads of:* A. Litoria nannotis; *B.* Limnodynastes ornatus; *C.* Litoria wotjulumensis. *From Tyler and Lungershausen (1986).*

---

[1] Robertson referred his specimens to '*U. rugosa*' but this was prior to the definitive studies of Davies and Littlejohn (1986) demonstrating that what was called '*rugosa*' was in reality a complex of species. What Robertson had was really *U. laevigata*.

PLATE 20 *Mismatch! A male* Litoria nasuta *with the wrong partner* (Notaden melanoscaphus). (M. Davies)

THE SEXUAL EMBRACE

There are two behavioural positions adopted by the coupling frogs, a condition described as being 'in amplexus'. In the Hylidae and the leptodactylid genus *Mixophyes*, the male rides upon the back of the female, grasping her above her armpits (axillary amplexus). In the remaining leptodactylid genera, and in the Microhylidae, Bufonidae and Ranidae the male remains behind the female and grasps her around the waist (inguinal amplexus).

To aid the grasp the sexually mature male is equipped with what are termed 'nuptial pads' on the inner surface of each thumb and sometimes on one or two adjacent fingers as well. To the eye each pad is a raised black structure. It may be in the form of a cluster of spines or extremely elaborate rosettes (Figure 28).

The nuptial pads develop at the start of the mating season, and it seems that the sites on which they appear are areas of tissue that are sensitive to elevated levels of circulating male sex hormones (Delrio *et al.*, 1980). After the mating season the nuptial pads are shed with the skin during the frequent habit of skin slough.

BODY LANGUAGE

In 1985, I accompanied an ABC film crew visiting the Eungella National Park near Mackay to film the then newly discovered gastric-brooding frog, *Rheobatrachus vitellinus*. Whilst searching for the frog in a mountain creek in dense rainforest, our attention was drawn to the strange behaviour of the little frog *Taudactylus eungellenis*. The frogs appeared to be waving to one another.

Perched upon a moss-covered boulder in the centre of the creek one of these pretty brown and yellow frogs gave every impression of attempting to draw attention to itself. Its antics were elaborate—first a series of tiny hops, each one propelling itself forward only a few millimetres; then a sequence of random movements of the right arm and leg, stretching and waving in an elaborate gymnastic display.

The object of the small frog's attention seemed to be another individual upon a boulder 1 m away in the stream. This frog was absolutely immobile. It was as animated as the rock upon which it stood! The active frog finally jumped into the water separating them, and quickly swam to the recipient's rock. Slowly it emerged and went up to the recipient and began to stroke its head and body. Eventually the activity stopped and the pair dived into the water and swam off together. Richard and James (1992) also reported seeing this behaviour.

Taudactylus eungellensis is one of those species that lack a vocal sac, so it seems likely that the behaviour observed was part of a courtship pattern that evolved in a habitat where the background noise level was so high that voice communication was ineffective.

There has been only one other report of similar frog body language. Those who watched the BBC TV documentary *The Living Planet* featuring Sir David Attenborough may remember a short sequence of a frog in Brunei waving its feet around. I was reminded of it when I witnessed the antics of *T. eungellensis*. (A paper by Harding (1982) describes the behaviour of the Brunei frog (*Staurois parvus*).)

PLATE 21 *Eggs of tiny* Cophixalus *sp. found beneath leaf litter in northern Queensland. (M. Trenerry)*

CHAPTER SEVEN

DEVELOPMENT

EGGS AND FERTILISATION

Whilst the male and female are coupled together (in amplexus) the mature ova are shed from the two lobes of the ovary and simply released into the body cavity. The eggs then enter funnels (ostia) leading to the left or right oviduct. Each oviduct is a long and elaborately convoluted tube partly lined with ciliated cells which, in conjunction with muscular contractions of the oviduct wall, drive the eggs down the tube and out through the cloaca. On their passage down the oviduct the eggs rotate and become coated with a thin layer of jelly.

Fertilisation takes place after the eggs have been shed, and involves elaborate movements by the male and female. However despite the frequency of this phenomenon, even in captive frogs there seem to be only two detailed accounts of the egg-laying behaviour: of *Geocrinia laevis* by Littlejohn and Martin (1964) and of *Litoria verreauxi* by Anstis (1976). In *G. laevis* Littlejohn and Martin recognised seven separate steps:

(1) The process was initiated by the female stretching forward and straightening her back.
(2) The male then gave 5–6 chirps. These were similar to the male warning call and the male tightened his grip with each chirp.
(3) The female began contractions in the lower abdomenal and cloacal region with about 30 contractions in all.
(4) During the ninth or tenth contraction the male arched forward, lowered the vent, and drew the legs up and out. Sperm were probably released at this stage. This position was held for about 20 seconds, being slowly released during the last 5 seconds.

(5) The female rapidly extruded about 8–10 eggs after the last contractions. These collected in the diamond-shaped area formed by her legs.
(6) Extrusion of eggs occurred at about 4 minute intervals, and each extrusion sequence lasted about 30 seconds.
(7) The eggs were left behind as the pair moved forward.

The account of Anstis (1976) described elaborate movements of the hindlegs of the male controlling the position of the extruded eggs. I have noticed fundamentally similar behaviour in *Limnodynastes tasmaniensis*: the male putting the tips of his toes or his feet together creating a dam, and preventing the eggs from floating to the surface. It is probably at such moments that he sheds the sperm upon them.

The act of fertilisation often is portrayed as one involving active sperm making their way to a passive egg. This may well describe events within the reproductive tract of mammals, but I think it is most unlikely in frogs. For example, in *L. caerulea* the female expels her eggs with such force that they pass through the sperm cloud and end up half a metre away. The sperm must almost be impaled upon the eggs! In other species observations suggest that the opportunity for possible contact of sperm and eggs is so brief that it may be normal for the eggs to bump into the sperm.

If eggs are removed from the female's body cavity prior to their passage down the oviducts, and then mixed with a sperm suspension, fertilisation does not take place. It appears that the jelly secreted around the eggs by the oviduct includes an ingredient essential for fertilisation. There is a good chance that this factor is prostaglandin E_2 which has been found in the oviducal secretions of frogs and in the reproductive tracts of other animals.

Following fertilisation (or perhaps whilst it is taking place) the jelly around the eggs absorbs water very rapidly and so swells. In the European frog *Rana temporaria* it has been found that the degree of enlargement is directly associated with the solutes in the water—the lower the salt concentration, the greater the distension of the jelly (Beattie, 1980).

Frogs' eggs vary considerably in size from a low of 0.8 mm diameter in *Litoria microbelos* to 5.3 mm in *Myobatrachus gouldii*. The size difference reflects the yolk reserves available to the developing embryo. Thus the larger the egg the longer the duration of independence from external food sources (and/or the larger the size of the emerging embryo).

The presence of black melanin pigment on the upper pole of eggs occurs in almost all species exposed to sunlight; it provides a filter which excludes the ultraviolet portion of the light spectrum which would be harmful to the developing embryo. Eggs laid in hidden situations have no need of, and so lack, the pigment and are just white or cream.

Published information on the eggs and tadpoles of Australian species is summarised in Table 6.

SPAWN

The term 'spawn' is applied here to the eggs and the jelly that surrounds them. It may be laid in a single mass, or separately in smaller clumps, or else the eggs may be laid individually. The various forms that the spawn can take in different genera are listed in Table 5.

In its simplest form the spawn consists of a large shapeless mass of eggs surrounded by jelly (Plate 22). Initially it may float upon the surface of the water and remain there. Subsequently it may sink to the floor, or it may fall to the floor as it is laid. The slightest deviation from this form is where the female attaches small portions of her eggs to a number of submerged grass stems in different parts of a pool, so distributing her total complement over a larger area. In captivity the large green tree frog *Litoria splendida* laid

spawn in an aquarium and, after depositing perhaps 200 eggs at a time, would attempt to escape, making numerous circuits of the aquarium—behaviour that in the wild would no doubt involve moving from one pool to another, spreading the eggs over a wide area. This would increase the chances of survival of at least some of the progeny both in terms of reducing dependence upon a single food source, and the unreliable nature of temporary pools.

To deposit loose clumps of eggs unattached, or weakly attached to vegetation is a strategy that is effective in static water, but for those species that breed in water that flows, even intermittently, the spawn must be secured more firmly or it will be washed away and, most likely, destroyed in the process. There is a trend in such species for the jelly surrounding the eggs to be much stronger with a consistency like gelatine. It is not known quite how this is achieved.

Amongst the various forms of spawn laid, one of the most common is, in fact, quite bizarre, and involves the construction of foam nests. It is characteristic of frogs of the genera *Adelotus, Heleioporus, Lechriodus, Limnodynastes, Megistolotis* and *Philoria*.

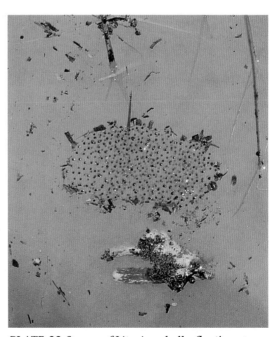

PLATE 22 *Spawn of* Litoria rubella *floating upon the surface of a temporary pool at Jabiru, NT.* (M.J. Tyler)

TABLE FIVE: Form of spawn in Australian frogs

GENUS	SPAWN	SPAWNING SITE	EXAMPLE	REFERENCE
Adelotus	foam nest	surface of water, usually hidden	*A. brevis*	Moore (1961)
Arenophryne	separate, unattached eggs	beneath surface of sand	*A. rotunda*	Roberts (1984)
Assa	liquefied jelly	surface of ground	*A. darlingtoni*	Ingram *et al.* (1975)
Bufo	long chain	in water	*B. marinus*	Breder (1946)
Cophixalus	single eggs	on soil beneath debris	*C. ornatus*	Zweifel (1985)
	chain	on soil	*Cophixalus* sp.	G. Werren (pers. comm.)
Crinia	small clumps	in water	*C. signifera*	Martin (1967a)
	film of eggs	undersurface of rocks at edge of creeks	*C. riparia*	Martin (1967a)
Cyclorana	large, irregular clump	in water	*C. australis*	Tyler *et al.* (1983
Geocrinia	loose, elongate	beneath moist litter on land	*G. rosea*	Main (1965)
Heleioporus	foam nest	in dry burrow near water; later flooded	*H. barycragus*[1]	Main (1965)
Kyarranus	large mass of jelly	on land in saturated soil	*K. sphagnicolus*	Littlejohn (1967)
Lechriodus	foam nest	surface of water	*L. fletcheri*	Moore (1961)
Limnodynastes	foam nest	surface of water exposed to sunlight	*L. tasmaniensis*[2]	Moore (1961)
Litoria	large clumps that sink to floor	in water	*L. caerulea*	Harrison (1922)
	small clumps attached to submerged vegetation	in water	*L. ewingi*	Waite (1929)
	film floating on surface	in water	*L. rubella*	Tyler *et al.* (1983)
	film attached to undersurface of rocks	in water	*L. meiriana*	Tyler *et al.* (1983)

[1] Reported as *H. australiacus*
[2] Roberts & Seymour (1989) report non-foamy nests in the south-east of South Australia

GENUS	SPAWN	SPAWNING SITE	EXAMPLE	REFERENCE
Litoria	single eggs or small clumps on floor	in water	*L. coplandi*	Tyler *et al.* (1983)
Megistolotis	foam nest	in water	*M. lignarius*	Tyler et al. (1979)
Mixophyes	cohesive clump	on land at edge of water	*M. fasciolatus*	Martin (1967a)
Myobatrachus	separate, unattached eggs	beneath surface of sand	*M. gouldii*	Roberts (1981)
Neobatrachus	long chain	in water	*N. wilsmorei*	Main (1965)
Notaden	long submerged chain	in water	*N. melano-scaphus*	Tyler et al. (1983)
Nyctimystes	cohesive clump attached to rock	in water	*N. dayi*	G. Ingram (pers. comm.)
Philoria	foam nest	saturated sphagnum	*P. frosti*	Littlejohn (1963)
Pseudophryne	separate eggs	beneath debris or in cavities out of water	*P. bibroni*	Woodruff (1976)
	separate eggs	in water	*P. douglasi*	Main (1964)
Rana	large clumps of loosely attached eggs	in water	*R. daemeli*	J. Menzies (pers. comm.)
Rheobatrachus	unknown	unknown, possibly on land	*R. silus*	—
Sphenophryne	single eggs	on soil beneath debris	*S. fryi*	Zweifel (1985)
Taudactylus	small clumps	under rocks in water	*T. eungellensis*	Liem & Hosmer (1973)
Uperoleia	small clumps	edge of pools	*U. inundata*	Tyler *et al.* (1983)

PLATE 23 *Foamy egg masses of* Limnodynastes dumerilli. *(A. Martin)*

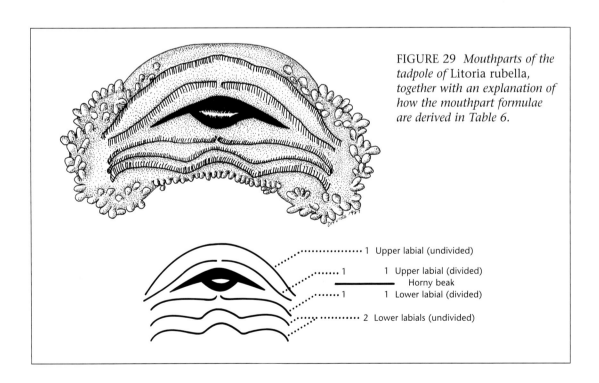

FIGURE 29 *Mouthparts of the tadpole of* Litoria rubella, *together with an explanation of how the mouthpart formulae are derived in Table 6.*

.... 1 Upper labial (undivided)

.... 1   1 Upper labial (divided)
————— Horny beak
.... 1   1 Lower labial (divided)

.... 2 Lower labials (undivided)

TABLE SIX: Published data on the eggs and tadpoles of Australian frogs

GENUS AND SPECIES	NO.OF EGGS	EGG DIAMETER (mm)	TADPOLE	TADPOLE TOOTH ROWS	MINIMUM TADPOLE LIFE (days)	REFERENCE
Adelotus brevis	270	1.7–1 8	lentic	$\dfrac{2\ \ 1\ \ 2}{1\ \ 2\ \ 1}$	71 @ *ca.* 20°C	Moore (1961) Watson & Martin (1973) Martin (1967a)
Arenophryne rotunda	6–11	5.5 ± 0.15 S.E.	within capsule	0	73	Roberts (1984)
Assa darlingtoni	10–11	2.2–2.6	in hip pouches of male	0	60	Straughan & Main (1966) Ingram, Anstis & Corben (1975) Ehmann & Swan (1985)
Bufo marinus	8000–25 000	1.7–2.0	lentic	$\dfrac{1\ \ 1\ \ 1}{3}$	30	Zug & Zug (1979) Breder (1946)
Crinia bilingua		1.1–1.2	lentic	$\dfrac{1\ \ 1\ \ 1}{1\ \ 2\ \ 1}$	13–14	Martin, Tyler & Davies (1980)
Crinia deserticola			lentic	$\dfrac{1\ \ 1\ \ 1}{1\ \ 2\ \ 1}$		Liem & Ingram (1977)
Crinia georgiana	*ca.* 70	1.7–2.8	lentic	$\dfrac{1\ \ 2\ \ 1}{1\ \ 2\ \ 1}$	35	Main (1957)
Crinia glauerti	*ca.* 70	1.0–1.3	lentic		130 @ 16°C	Main (1957)
Crinia haswelli		*ca.* 2.0	lentic	$\dfrac{1\ \ 1\ \ 1}{2}$		Watson & Martin (1973)
Crinia insignifera	66–268	1.1–1.5	lentic		*ca.* 60	Main (1957, 1965)
Crinia parinsignifera		1.2–1.7	lentic	$\dfrac{1\ \ 1\ \ 1}{1\ \ 2\ \ 1}$	79 @ 6–18°C	Straughan & Main (1966) Watson & Martin (1973)
Crinia pseudinsignifera	77–207	1.2–1.5	lentic	$\dfrac{1\ \ 1\ \ 1}{2}$	82	Main (1957)
Crinia riparia		1.96	lotic	$\dfrac{2}{1\ \ 2\ \ 1}$		Littlejohn & Martin (1965) Watson & Martin (1973)
Crinia signifera	100–150	1.3–1.6	lentic	$\dfrac{1\ \ 1\ \ 1}{1\ \ 2\ \ 1}$	49 @ 14–15°C	Straughan & Main (1966) Moore (1961) Watson & Martin (1973) Gollman (1991)
Crinia tasmaniensis	46–69	2.0 – 2.27	lentic	$\dfrac{1\ \ 1\ \ 1}{1\ \ 2\ \ 1}$	3–4 months	Martin (1965) Blanchard (1929) Watson & Martin (1973) Martin & Littlejohn (1982)
Crinia tinnula	33–118	1.1 – 1.2	lentic			Straughan & Main (1966)

GENUS AND SPECIES	NO. OF EGGS	EGG DIAMETER (mm)	TADPOLE	TADPOLE TOOTH ROWS	MINIMUM TADPOLE LIFE (days)	REFERENCE
Cophixalus infacetus	8 – 11		intra-capsular	0		Zweifel (1985)
Cophixalus neglectus	14		intra-capsular	0		Zweifel (1985)
Cophixalus ornatus	22		intra-capsular	0		Zweifel (1985)
Cyclorana australia	1000–7000	1.5 – 1.8	lentic	$\frac{1\ 1\ 1}{1\ 2\ 1}$	29	Tyler, Crook Davies (1983)
Cyclorana brevipes			lentic	$\frac{1\ 1\ 1}{1\ 2\ 1}$	25	Richards & Alford (1993)
Cyclorana cryptotis			lentic	$\frac{1\ 1\ 1}{1\ 2\ 1}$	24 @ 28°C	Tyler, Davies & Martin (1982)
Cyclorana longipes	1000–1615	1.2 – 1.5	lentic	$\frac{1\ 1\ 1}{1\ 2\ 1}$	33 @ 30°C	Tyler, Crook & Davies (1983)
Cyclorana platycephala			lentic	$\frac{1\ 1\ 1}{1\ 2\ 1}$	30	Spencer (1896) Main (1968) Watson & Martin (1973)
Geocrinia laevis	70–150	2.9 – 3.3	intra-capsular (in part) lentic	$\frac{1\ 1\ 1}{1\ 2\ 1}$		Martin & Littlejohn (1982) Watson & Martin (1973) Littlejohn & Martin (1964) Gollman & Gollman (1991)
Ceocrinia leai	52–96	1.7–2.0			>120	Main (1957, 1968)
Geocrinia rosea	26–32	2.35		0	60–70	Main (1957, 1968)
Geocrinia victoriana	90–162	2.9 – 3.3	intra-capsular (in part) lentic	$\frac{1\ 1\ 1}{3}$	6–8 months	Littlejohn & Martin (1964), Gollman & Gollman (1991)
Heleioporus albopunctatus	250–700		lentic	$\frac{5\ 1\ 5}{1\ 2\ 1}$		Lee (1957) Davies (1991)
Heleioporus australiacus	775–1239		lentic	$\frac{5\ 1\ 5}{1\ 2\ 1}$		Watson & Martin (1973)
Heleioporus barycragus			lentic	$\frac{3\ 2\ 3}{1\ 2\ 1}$		Lee (1967)
Heleioporus eyrei	80–500	3.3	lentic	$\frac{4\ 1\ 4}{1\ 2\ 1}$		Main (1965)
Heleioporus inornatus	100–250		lentic	$\frac{4\ 1\ 4}{1\ 2\ 1}$		Main (1965) Lee (1967)
Heleioporus psammophilus		3.5	lentic	$\frac{4\ 1\ 4}{1\ 2\ 1}$		Main (1965)

GENUS AND SPECIES	NO. OF EGGS	EGG DIAMETER (mm)	TADPOLE	TADPOLE TOOTH ROWS	MINIMUM TADPOLE LIFE (days)	REFERENCE
Kyarranus kundagungan		*ca.* 3.1		0		Ingram & Corgan (1875)
Kyarranus loveridgei	20 – 30		in liquefied jelly nest	0		Moore (1961)
Kyarranus sphagnicolus	44 – 52	2.9 – 3.2		0	1 month	Moore (1958) Anstis (1981) Watson & Martin (1973)
Lechriodus fletcheri	300	1.7	lentic	$\frac{5\ 1\ 5}{1\ 2\ 1}$		Watson & Martin (1973) Martin (1967a)
Limnodynastes dorsalis			lentic			
Limnodynastes dumerili	3900	1.7	lentic	$\frac{4\ 1\ 4}{1\ 2\ 1}$	12–15 months	Martin (1965)[1] Martin (1967a) Martin & Littlejohn (1982) Davies (1992)
Limnodynastes fletcheri		1.3–1.6	lentic	$\frac{1\ 2\ 1}{1\ 2\ 1}$	60 @ 30°C	
Limnodynastes interioris		1.7	lentic	$\frac{5\ 1\ 5}{1\ 2\ 1}$		Watson & Martin (1973)
Limnodynastes ornatus	1000–1630	1.0 – 1.3	lentic	$\frac{3\ 1\ 3}{1\ 2\ 1}$	21	Tyler, Crook & Davies (1983)
Limnodynastes peronii	705–1009	1.5	lentic	$\frac{2\ 2\ 2}{1\ 2\ 1}$	11–12	Martin & Littlejohn (1982) Moore (1961)
Limnodynastes spenceri	1128	1.5	lentic	$\frac{1\ 1\ 1}{1\ 2\ 1}$	40	Main Calaby (1957) Spencer (1896) Main (1968)
Limnodynastes tasmaniensis	88–1359	1.1–1.4	lentic	$\frac{4\ 1\ 4}{1\ 2\ 1}$	3–5 months	Martin (1965) Horton (1982b) Martin & Littlejohn (1982)
Limnodynastes terraereginae			lentic	$\frac{4\ 1\ 4}{1\ 2\ 1}$	71 @ 30°C	Davies (1992)
Litoria bicolor		1.9–2.9	lentic	$\frac{2}{3}$	77 @ 30°C	Tyler, Crook & Davies (1983)
Litoria booroolongensis			lotic	$\frac{2}{3}$		Anstis (1974)
Litoria burrowsae			lentic	$\frac{1\ 1\ 1}{1\ 2\ 1}$		Martin (1967b)
Litoria caerulea	2000–3000	1.1–1.4	lentic	$\frac{2\quad 2}{3}$	38 @ 30°C	Tyler, Crook & Davies (1983) Moore (1991)
Litoria chloris			lentic	$\frac{1\ 1\ 1}{1\ 2\ 1}$	41 @ 27°C	Watson & Martin (1979)

[1] as *L. doralis*

GENUS AND SPECIES	NO. OF EGGS	EGG DIAMETER (mm)	TADPOLE	TADPOLE TOOTH ROWS	MINIMUM TADPOLE LIFE (days)	REFERENCE
Litoria citropa	890	1.7–1.8	lentic	$\frac{1\ 1\ 1}{1\ 2\ 1}$	2 months	Tyler & Anstis (1975)
Litoria cooloolensis				$\frac{1\ 1\ 1}{3}$		Liem (1974b)
Litona coplandi			lotic	$\frac{2}{3}$	52	Tyler, Crook & Davies (1983
Litoria dahlii			lentic	$\frac{2}{3}$		Tyler, Crook & Davies (1983)
Litoria ewingi			lentic	$\frac{1\ 1\ 1}{1\ 2\ 1}$	6–7 months	Martin & Watson (1971) Martin & Littlejohn (1982)
Litoria fallax	263		lentic	$\frac{1\ 1\ 1}{3}$	118 @ 20°C	Watson & Martin (1979)
Litoria genimaculata	843	2.1–2.6	lotic	$\frac{1\ 1\ 1}{1\ 2\ 1}$		Davies (1989)[2]
Litoria gracilenta			lentic	$\frac{1\ 1\ 1}{1\ 2\ 1}$	112 @ 20°C	Watson & Martin (1979)
Litoria inermis	96–330	1.3–1.4	lentic	$\frac{1\ 2\ 1}{2}$	74 @ 30°C	Tyler, Crook & Davies (1983) Davies, Martin & Watson (1983)
Litoria infrafrenata	ca. 430	1.9	lentic	$\frac{1\ 1\ 1}{3}$	58 @ 24–28°C	Banks *et al.* (1983)
Litoria lesueuri		1.5	lotic	$\frac{2}{3}$		Martin, Littlejohn & Rawlinson (1966)
Litoria longirostris	29–60	2.0–2.4				McDonald & Storch (1993)
Litoria meiriana	39–115		lotic	$\frac{1\ 2\ 1}{2}$	30	Tyler, Crook & Davies (1983)
Litoria microbelos	60	0.8 – 0.9	lentic			Tyler, Crook & Davies (1983)
Litoria nannotis		2.7 – 3.4	lotic	$\frac{2}{3}$		Liem (1974a)
Litoria nasuta	50–100	1.0–1.3	lentic	$\frac{2}{3}$	31 @ 30°C	Tyler, Crook & Davies (1983
Litoria nigrofrenata			lentic	$\frac{1\ 1\ 1}{3}$		Crossland & Richards (1993)
Litoria nyakalensis			lotic	$\frac{2}{3}$		Richards (1992)
Litoria pallida	50–350	1.2–1.4	lentic	$\frac{2}{3}$	54	Tyler, Crook & Davies (1983) Davies, Martin & Watson (1983)

[2] as *L. eucnemis*

GENUS AND SPECIES	NO.OF EGGS	EGG DIAMETER (mm)	TADPOLE	TADPOLE TOOTH ROWS	MINIMUM TADPOLE LIFE (days)	REFERENCE
Litoria pearsoniana		3.0–3.5	lentic	$\frac{1\ \ 1\ \ 1}{1\ \ 2\ \ 1}$	2–2.5 months	McDonald & Davies (1990)
Litoria peronii		1.5–1.6	lentic	$\frac{1\ \ 1\ \ 1}{1\ \ 2\ \ 1}$		Martin & Watson (1971) Martin *et al.* (1979)
Litoria personata			lotic	$\frac{1\ \ 1\ \ 1}{2\ \ 1\ \ 2}$		Tyler, Davies & Martin (1978)
Litoria phyllochroa			lentic	$\frac{1\ \ 1\ \ 1}{1\ \ 2\ \ 1}$		Martin & Watson (1971)
Litoria raniformis			lentic	$\frac{1\ \ 1\ \ 1}{1\ \ 2\ \ 1}$		Martin & Littlejohn (1982)
Litoria rheocola		1.4–1.2	lotic	$\frac{2}{3}$		Liem (1974a), Richards (1992)
Litoria rothii	504	1.3–1.4	lentic	$\frac{1\ \ 1\ \ 1}{3}$	65	Tyler, Crook & Davies (1983)
Litoria rubella	300–715	1.0–1.1	lentic	$\frac{1\ \ 1\ \ 1}{1\ \ 2\ \ 1}$	37	Tyler, Crook & Davies (1983)
Litoria subglandulosa		2.1	lotic	0		Tyler & Anstis (1975)
Litoria tornieri	69–190	1.4	lentic		46 @ 30°C	Tyler, Crook & Davies (1983) Davies, Martin & Watson (1983) Davies
Litoria tyleri		1.3–1.4	lentic	$\frac{1\ \ 1\ \ 1}{1\ \ 2\ \ 1}$		Martin *et al.* (1979)
Literia verreauxi	522–1011	1.2	lentic	$\frac{1\ \ 1\ \ 1}{1\ \ 2\ \ 1}$	29	Martin (1965) Anstis (1976)
Litoria wotjulumensis	200	1.6–1.9	lentic	$\frac{1\ \ 1\ \ 1}{3}$	39 @ 30°C	Tyler, Crook & Davies (1983)
Litoria xanthomera			lentic	$\frac{1\ \ 1\ \ 1}{1\ \ 2\ \ 1}$		Davies, McDonald & Adams (1986)
Megistolotis lignarius	352	1.8–1.9	lotic	$\frac{4\ \ 2\ \ 4}{3}$	180	Tyler, Davies & Martin (1982)
Mixophyes balbus		2.8	lotic	$\frac{10\ \ 1\ \ 10}{1\ \ 2\ \ 1}$		Watson & Martin (1973)
Mixophyes schevilli			lotic	$\dfrac{5\ \ \ 1\ \ \ 5}{\dfrac{5\text{–}6\ \ \ \ 5\text{–}6}{1\ \ 2\ \ 1}}$		Davies (1991)
Myobatrachus gouldii	23–38	5.3	intra-capsular	0		Roberts (1981)
Neobatrachus aquilonius	1426	1.9				Tyler, Davies & Martin (1982)

GENUS AND SPECIES	NO.OF EGGS	EGG DIAMETER (mm)	TADPOLE	TADPOLE TOOTH ROWS	MINIMUM TADPOLE LIFE (days)	REFERENCE
Neobatrachus centralis			lentic	$\frac{2\ 2\ 2}{1\ 2\ 1}$		Davies (1991)
Neobatrachus kunapalari			lentic	$\frac{1\ 2\ 1}{1\ 2\ 1}$		Davies (1991)
Neobatrachus pictus		2.2	lotic	$\frac{2\ 1\ 2}{1\ 2\ 1}$		Watson & Martin (1973)
Neobatrachus sudelli			lentic	$\frac{2\ 1\ 2}{1\ 2\ 1}$	4.5–7 months	Martin (1965)[3]
Neobatrachus wilsmorei			lentic	$\frac{2\ 1\ 2}{1\ 2\ 1}$		Davies (1991)
Notaden melanoscaphus	50–1200	1.3–1.6	lotic	$\frac{1\ 1\ 1}{1\ 1\ 1}$	53	Tyler, Crook & Davies (1983)
Notaden nichollsi				$\frac{2\ 1\ 2}{1\ 2\ 1}$	16	Main (1968)
Nyctimystes dayi	107	2.3–2.6	lotic	$\frac{2}{3}$	90–120	Davies & Richards (1990)
Philoria frosti	81–110	3.9	in liquefied jelly nest	0	ca. 40	Littlejohn (1963) Watson & Martin (1973)
Pseudophyrne australis	15–20	2.6–2.5	lentic		28–40	Moore (1961) Penggilley (1973)
Pseudophyrne bibroni	77–190	2.0–2.5	intra-capsular (in part) lentic	$\frac{1\ 1\ 1}{1\ 2\ 1}$	120–210	Martin (1965) Woodruff (1976)
Pseudophyrne corroboree	12–30	*ca.* 3	lentic	$\frac{1\ 1\ 1}{1\ 2\ 1}$	180–210	Watson & Martin (1973) Pengilley (1973)
Pseudophryne dendyi	75–164	2.6–3.2	lentic	$\frac{1\ 1\ 1}{3}$		Moore (1961) Woodruff (1976)
Pseudophryne douglasi	89	2.0–2.3	lentic	$\frac{1\ 1\ 1}{1\ 2\ 1}$	90–120	Main (1964)
Pseudophryne semimarmorata	119–169	2.1–2.5	intra-capsular (in part) lentic			Martin (1965) Woodruff (1976)
Rana daemeli			lentic	$\frac{1\ 1\ 1}{1\ 2\ 1}$		Richards (1992)
Rheobatrachus silus	26–40	4.7	in stomach of female	0	*ca.*6 weeks	Horton & Tyler (1982) Tyler & Davies (1983)
Rheobatrachus vitellinus	22		in stomach of female	0		McDonald & Tyler (1984)

[3] as *N. pictus*

GENUS AND SPECIES	NO. OF EGGS	EGG DIAMETER (mm)	TADPOLE	TADPOLE TOOTH ROWS	MINIMUM TADPOLE LIFE (days)	REFERENCE
Sphenophryne fryi	7–12		intra-capsular	0		Zweifel (1985)
Taudactylus acutirostris			lotic	$\frac{1\ 1\ 1}{1\ 2\ 1}$		Liem & Hosmer (1973)
Taudactylus diurnus		2.2	lotic	0		Watson & Martin (1973)
Taudactylus eungellensis	30–50	2.2–2.6	lotic	0		Liem & Hosmer (1973)
Taudactylus liemi	34–51	1.7–2.5				Ingram (1980)
Taudactylus rheophilus	35–50	1.8–2.4				Liem & Hosmer (1973)
Uperoleia inundata	400	1.1–1.4	lentic	$\frac{1\ 1\ 1}{3}$	51	Tyler, Crook & Davies (1983)
Uperoleia laevigata[4]		1.3	lentic	$\frac{1}{3}$		Moore (1961)
Uperoleia lithomoda			lentic	$\frac{1\ 1\ 1}{3}$		Davies, McDonald & Corben (1986)
Uperoleia mimula			lentic	$\frac{1\ 1\ 1}{3}$		Richards & Alford (1993)
Uperoleia tyleri [4]		1.5	lentic	$\frac{1\ 1\ 1}{1\ 2\ 1}$		Watson & Martin (1973)

[4] as *U. marmorata*

Foam nests are constructed by frogs on different continents (Martin, 1970) but the techniques that they employ vary. In the Australian species so far observed, the female paddles, raising her hands to the surface of the water, so trapping small air bubbles which she then throws in a downward and backward motion beneath her body. The bubbles bounce upon her abdomen, and pass between her legs, where they become trapped in the eggs and jelly emerging from her cloaca. There is nothing haphazard about the activity (Parker, 1940; Tyler and Davies, 1979).

Photographs of foam nests of *Physalaemus pustulosus* of Panama and *Chiromantis petersi* of Kenya (Duellman and Trueb, 1986) show that the male beats up the extruding jelly with his hind legs and, as far as it is possible to judge from illustrations, the foam appears to have a different consistency (perhaps more cohesive) than the foam nests constructed by Australian species. There is, for example, no way in which a foam nest of an Australian species would maintain its integrity suspended above the ground in the way that foreign species do.

Irrespective of the nature of the foam nest, the functional significance of foam nesting probably also varies. For example, *C. petersi* of Africa, which constructs the nest in trees, appears to do so in order to provide a moist environment out of water. In the case of *Philoria* and *Kyarranus*, which construct nests in soggy ground or sphagnum, the nest may well have a similar function. Similarly *Heleioporus* lays eggs in a foam nest in a burrow prior to flooding. But *Adelotus*, *Lechriodus*, *Limnodynastes* and *Megistolotis* all construct their nests in water where the spawn forms a raft from which the eggs hang suspended just below the surface. In the cooler, temperate parts of Australia this habit maintains the eggs at the surface where, if there is any thermal stratification, water temperatures will be highest. Because the rate of development is related to temperature, it may serve to accelerate early development, an action that, in the case of a mixed community breeding simultaneously, could provide the tadpoles of foam nesters with an advantage.

The foam nest of *Limnodynastes ornatus* of northern and eastern Australia is unusual in that it collapses rapidly and spreads out to form a film upon the surface of the water. This result does not imply any difference in the nature of the foam, but is a consequence of its inability to maintain its physical structure in higher water temperatures. This has been demonstrated by Morisson (1982) who showed that *L. tasmaniensis* spawn collapsed to form a similar surface film when transferred from the water in which it had been laid at 20°C to water at 30°C.

About one-quarter of the Australian frog species lay eggs out of water and either pass their entire developmental period on land (direct development), or else enter water at a relatively advanced stage of development (delayed emergence).

The habits of direct development and delayed emergence are associated with an increase in the size of the individual eggs, a reduction in the number of eggs laid and, often, with adult (assumed parental) attendance during the period of terrestrial development. In addition (although this is not exclusive to such species) there is a tendency for the egg clutch to be laid in a hidden site—for example, beneath debris or in the soil, and also for the outer capsule around the eggs to be tough and resilient.

PLATE 24 Pseudophryne major *with eggs. (M. Trenerry)*

PLATE 25 *Pairs of* Litoria gracilenta, *from north Queensland. (M. Trenerry)*

PLATE 26 Litoria serrata, *a pool dwelling tadpole. (M. Trenerry)*

Delayed emergence is characteristic of the toadlets (*Pseudophryne* species) which lay eggs in cavities in the ground or beneath leaf litter in autumn, and which hatch in winter following flooding of the deposition site. The bizarre always attracts the most interest. For example Bradford and Seymour (1985) compared oxygen consumption of tadpoles of *P. bibroni* in situations simulating early hatching or delayed emergence, by varying the exposure of tadpoles to flooding. They found that in those individuals in which the release from the egg capsules was delayed, oxygen consumption was lowered, permitting the dependence upon yolk reserves to be extended for up to one 100 days.

The accumulation of waste products also presents a major problem to embryos remaining within egg capsules. Domm and Janssens (1971) examined *P. corroboree* and concluded that no nitrogen is excreted by that species prior to hatching. It was only after hatching that ammonia was excreted and nitrogen levels fell.

Direct development represents the ultimate step in the elimination of dependence upon free bodies of water. However this form of development is only possible in situations where the substrate with which the eggs are in contact is saturated. Thus, the availability of water is guaranteed.

Amongst Australian species, 40 is the maximum number of eggs laid by a species exhibiting direct development.

Parental Care

The transportation of tadpoles in hip pockets by *Assa darlingtoni*, and gastric brooding by the two *Rheobatrachus* species represent genuine and highly bizarre examples of parental care. These examples are discussed in Chapters 3 and 12. However there remain several instances in which one parent (or assumed parent) remains with the eggs until the stage of hatching. What role the adults play is unknown, and it may be more accurate to describe this phenomenon as 'attendance' rather than parental care.

Adult attendance occurs in several genera. In *Adelotus brevis* a male usually remains beneath or directly adjacent to the floating spawn clump. In *Pseudophryne* species either a male or a female remains with (commonly upon) the terrestrial eggs, whilst in the microhylid genera *Cophixalus* and *Sphenophryne*, adults remain very close to the egg chains.

Little is known of predation upon spawn and, without knowing the predators, it is not possible to say whether the adult frogs in attendance could, in any way, prevent such predation.

It is becoming apparent that many granular glands in frog skin have antifungal and antibacterial activity, hence it is possible that contact with frog skin may impart to terrestrial eggs antibiotic protection.

Recently it has been reported that in northern Queensland *Litoria lesueuri* created a 'nest'

in the form of a depression on exposed, sandy creek banks. The nests are roughly 250 mm in diameter and 50 mm deep, and the spawn is laid within them (Richards and Alford, 1992; Richards, 1993).

Tadpoles

The development of any tadpole can be split up into a series of stages according to a scheme proposed by Gosner (1960). The stages in the development of *Litoria rubella* are shown in Figure 30.

Without holding too biased an opinion, I regard the transformation of a small bag of muscles and intestines, called a tadpole, into a handsome little frog, a matter for awe and wonder. The frequency with which I watch it happen does nothing to dim the drama of the event. Perhaps this is so simply because this change of state is such a radical step in which, with the exception of the limbs, every part of the internal and external structure of the body undergoes a metamorphosis. In fact it is the development of the limbs that provides the only continuity of form between tadpole and frog.

The word 'tadpole' is an extremely old one. It is derived from the medieval English 'taddepolle' or 'taddepol'. 'Tadde' means toad and 'poll' head—literally a toad that seems to be all head. It is certainly appropriate for the tadpoles of the true toads of the genus *Bufo* which do seem to have a disproportionately large head in relation to the size of their bodies, but not for other species.

Tadpoles differ from species to species much less than the frogs that they ultimately change into. Whereas frogs can be distinguished by differences in the colour and patterns upon their skin, most tadpoles tend to be brown or black, and distinct markings are exhibited by only a few. The most striking differences that exist involve proportions of the length of the tail against the body, the depth of the fins and the nature of the structures around the mouth. But in situations where adults of different species are closely related and distinguished in their structure only by minor features of size or markings, their respective tadpoles usually are indistinguishable. This evolutionary conservatism probably reflects the fact that there is limited need (and options) for diversity in the relatively similar aquatic envir-onments which they share.

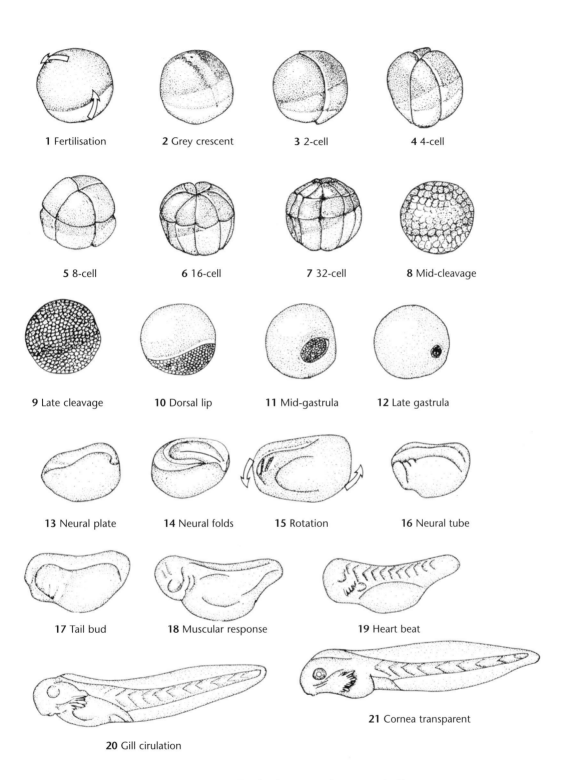

1 Fertilisation **2** Grey crescent **3** 2-cell **4** 4-cell

5 8-cell **6** 16-cell **7** 32-cell **8** Mid-cleavage

9 Late cleavage **10** Dorsal lip **11** Mid-gastrula **12** Late gastrula

13 Neural plate **14** Neural folds **15** Rotation **16** Neural tube

17 Tail bud **18** Muscular response **19** Heart beat

21 Cornea transparent

20 Gill cirulation

FIGURE 30 *Stages in the development of* Litoria rubella.

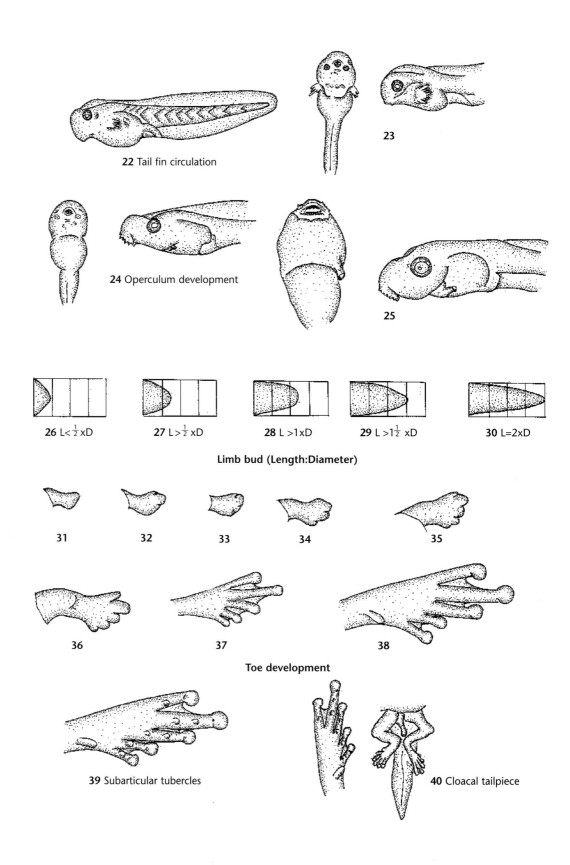

22 Tail fin circulation

23

24 Operculum development

25

26 L < ½ xD   **27** L > ½ xD   **28** L > 1xD   **29** L > 1½ xD   **30** L = 2xD

Limb bud (Length:Diameter)

31   **32**   **33**   **34**   **35**

36   **37**   **38**

Toe development

39 Subarticular tubercles   **40** Cloacal tailpiece

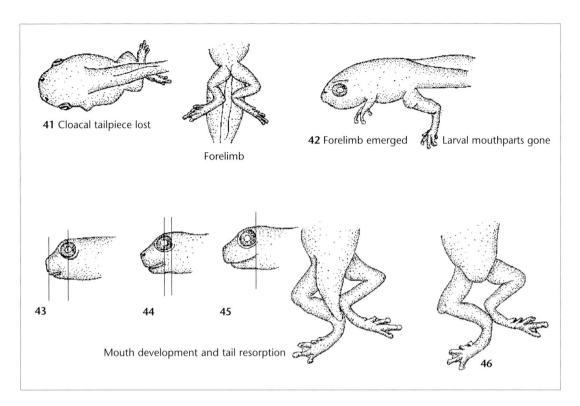

41 Cloacal tailpiece lost

Forelimb

42 Forelimb emerged   Larval mouthparts gone

43   **44**   **45**

Mouth development and tail resorption

46

The tadpole is a free-living creature in water. The early part of its development, prior to its emergence from the outer capsule that surrounds it, is commonly referred to as an embryo. Problems in terminology arise because a significant proportion of the world's species (perhaps 15 per cent), do not have free-living tadpoles and instead pass their entire period of development within the capsule that surrounds the egg. Thus if we try to subdivide tadpoles into meaningful groups, the first involves this fundamental issue of their life-style: free-living or intra-capsular. Inevitably the ova of those species that have intracapsular development are large

(generally greater than 3.0 mm in diameter), to provide sufficient yolk to nourish the embryo.

Amongst those species that have a free-living tadpole, development follows a sequence of events termed 'stages' (Figure 30). However, two kinds of tadpoles may be recognised, depending upon whether they live in, and are adapted to, static (lentic) or to flowing (lotic) water. The species living in lentic environments are those that are considered most 'typical', with bulbous bodies, mouths at the anterior end of the head and deep fins, whereas lotic species tend to have flattened bodies, and a modified head in which the mouth is sucker-like and on the undersurface

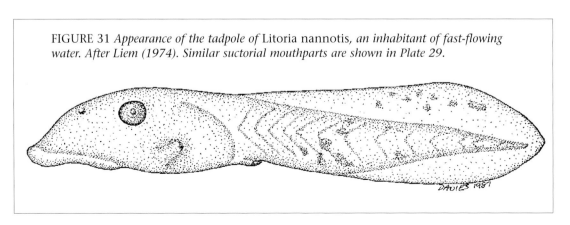

FIGURE 31 *Appearance of the tadpole of* Litoria nannotis, *an inhabitant of fast-flowing water. After Liem (1974). Similar suctorial mouthparts are shown in Plate 29.*

of the head to permit attachment to rocks, and narrow fins (Figure 31). However the ability to attach to submerged objects is not confined to species with sucker mouths, for Gradwell (1975) found that tadpoles of the lentic *Pseudophryne bibroni* could hold onto the inner surfaces of a glass aquarium. He believed that they did so by hooking their teeth into algae coating the glass.

The mouth of tadpoles is quite unlike any other vertebrate mouth. It consists of a slightly funnelled disc surrounded partly or entirely by papillae. Upon the disc are a series of rows of black, labial teeth resembling a comb, surrounding two, black, crescentic structures called the upper and lower horny beak (Figure 29).

Most tadpole feeding involves browsing. What a tadpole does is to open the mouth wide, so bringing the labial teeth forwards; the lips are then pulled slightly towards one another, scraping the teeth against the food material and shredding off small fragments in the process. The two halves of the horny beak overlap one another and can shear large particles into small portions. However, it would

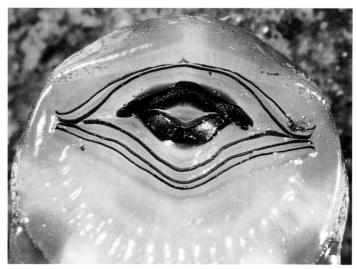

PLATE 27 (Left): *Spawn of* Notaden melanoscaphus. *(G.A. Crook)*

PLATE 28 (Top): *Pair of* Litoria lesueuri. *(M. Trenerry)*

PLATE 29 (Above): *Mouthparts of tadpole of* Nyctimystes dayi.
(M. Trenerry)

be wrong to view tadpoles as exclusively small particle feeders, for I have found a large, intact caterpillar and long strands of vegetation in the gut of *Neobatrachus pictus* (Tyler, 1976b).

As well as actively foraging for food, there is also a passive process by which additional items are ingested. Cords of mucus are secreted within the internal gills and pass backwards to the oesophagus, and these cords act as an entrapment mechanism. Included within the water passing in this passage through the mouth, across the gill arches and out through a short tube termed the spiracle (as a part of the process of respiration), are bacteria, foramanifera, protozoans and other forms of microscopic life. These become trapped in the mucus cords and can constitute a significant component of the dietary intake. Food capture is therefore a fairly complex, but effective, system of scraping, chopping and extracting.

The greater part of the tadpole's body is taken up by the long, coiled intestines commonly visible through the body wall. The intestine is always so crammed with food material that it can be viewed as a conveyor belt rapidly transporting food from the mouth. The actual length of the intestines has been found to vary with the diet: experiments have shown that those fed on meat develop shorter intestines than those fed on plant material. There is no clearly distinguishable stomach but instead a short and slightly modified region of gut termed the manicotto glandulare which is thought to secrete enzymes aiding digestion. In contrast with the adult stomach there is no evidence that the manicotto secretes hydrochloric acid so vital for the breakdown of food. However the evolution of the use of hydrochloric acid in vertebrate digestion is thought primarily to have been the need for the inhibition of putrefaction of food within the body. It may be that the passage of food through the tadpole intestine is so rapid that there is not sufficient time for putrefaction to occur.

The tadpole tail is generally two to three times the length of the body. Along its length lies a series of V-shaped muscle blocks supporting a dorsal and a ventral fin varying in depth. Because the tail is particularly thin it can be bent to a remarkable degree, and the passage of the tadpole through the water is accomplished by an undulating flapping movement. Thus, whereas a fish moves through water smoothly, the tadpole is more snake-like, swinging its tail from side to side, with the amplitude and frequency of the swinging determining the speed of movement. When it wishes to travel very slowly only the tip of the tail is moved; when travelling rapidly the entire tail is involved.

I have never encountered a tadpole in a solitary state. This is not that surprising because frogs lay varying numbers of eggs together and so inevitably tadpoles will share a resource. And because each has an independent life there will be some degree of competition for that resource. At one extreme is a vast volume of water with unlimited food supply and a tadpole density sufficiently low to make competition negligible. At the other extreme is a small volume of water, limited food resources and a high density of tadpoles. Both of these extremes occur in nature with a tendency for northern ephemeral pools to be less favourable.

When frogs lay spawn they do not seem to be assessing whether the water in which they deposit it will persist for long enough for the tadpoles to have time to complete metamorphosis. Thus the progeny of many clutches of spawn will simply dry up. However in eastern Australia there is a species whose tadpoles frequently seem to survive by eating each other. Tadpoles certainly do eat each other in any community, but not to the extent witnessed in *Lechriodus fletcheri*. Martin (1975) used the term 'obligate cannibalism' to describe the phenomenon. What happens is that the spawn is laid in a foam nest in a small quantity of water. The tadpoles hatch and firstly eat all of the plant and animal material there. When these resources are exhausted they eat each other and the fittest (perhaps the most agile) survive. In such a situation individuals with the most rapid growth rates will be at an advantage. By being bigger they may constitute larger targets, but they have bigger jaws and no doubt can do much more damage to smaller individuals than vice versa.

A fascinating aspect is that the density of tadpoles in a given volume of water influences their rate of growth. Even in the presence of abundant food, large numbers together tend to metamorphose at a smaller size (and take longer to complete development). There has been a number of experiments performed to elucidate this phenomenon, one of the earliest of which was that of Richards (1962); in Australia the only study seems to be that of Sokol (1984) who used the tree frog *Litoria ewingi* as a subject. Sokol noted that when tadpoles were maintained in dense groups they had a slower growth rate, longer larval periods and a smaller size at metamorphosis than tadpoles reared at a lower density. Confirming the observations of Richards (1962) he demonstrated that these effects were due to unknown substances released by the tadpoles rather than behavioural interaction between the tadpoles.

In the light of the speculation of Wassersug and Karmazyn (1984) that prostaglandin E_2 secreted by individual tadpoles inhibits their stomach development (this stops when the tadpole turns into a frog), it is perfectly possible, in a dense culture, for PGE_2 secretion into the water to be so high that the concentration could have an effect upon the entire colony. An individual tadpole might reach the stage of switching off PGE_2 release but continue to have its development inhibited by the PGE_2 released into the water by less developed siblings. The necessary experiments to test this hypothesis have yet to be performed.

It is probably rare for a species to have exclusive use of a particular breeding site. More likely there will be anything from three or four to a dozen species sharing the one resource. Just how tadpoles of these individual species interact with one another is understood poorly. It is recognised that they may feed in different areas and on different material, so using different components of the resource, but there are few data on this aspect of tadpole life.

In Australia there have been only two published studies of the interactions of tadpoles of different species, and each has involved closely related taxa: an examination of *Pseudophryne bibroni* and *P. semimarmorata* by Wiltshire and Bull (1977) and of *Crinia riparia* and *C. signifera* by Odendaal and Bull (1980). Each of these studies involved pairs of similar species that have minimal overlap in their distributions, hence raising the question of which behavioural or other trait was responsible for maintaining that separation: why couldn't one species occupy the area in which the other occurred? Was the boundary between the species stable?

The aggregation of tadpoles into schools, much like fish, is a common phenomenon in northern Australia and appears to indicate social interaction. Fundamentally three forms of schooling can be observed. In shallow water (perhaps no more than 2 or 3 cm deep), tadpoles may be gathered together so that only a small proportion of the pool is occupied. In general this action appears to be associated with the mutual selection of the cooler rather than the warmer portion of the pool. In shallow water the difference in temperature may be only slight (a matter of one or two degrees Celsius). Generally it is presumed that the selection of a cooler zone is associated with higher levels of oxygen in the cooler water.

The second form of schooling takes place in deeper water at least 0.3 m deep, (and generally more) where a three-dimensional mass of tadpoles hangs suspended, extending from surface to floor. This is extremely common in *Cyclorana australis*, in which individuals in the school seem to constantly change position. In such a school the motivation is suggested to be one of protection, where the chances of predation upon individuals remaining within this school will be reduced. Such schools form and reform by day but disperse to the edges of the pools at night.

Finally there are the daytime congregations of high numbers at the periphery of pools (Plate 30) observed on the flood plains of the Alligator River in the Northern Territory. For such a phenomenon neither a temperature gradient nor protection from predation are likely answers. If anything, the existence of what are clearly hundreds of thousands of tadpoles in one spot would constitute rich and easy pickings for wading birds.

But the highest death rate amongst summer breeders is the failure of the pool in which eggs are deposited, to persist for long enough for them to complete development through to metamorphosis.

METAMORPHOSIS

The process of the climax of metamorphosis commences when the tadpole thrusts one arm out through the spiracle and the other arm through a weakened area on the other side of the branchial chamber. The blocking of the spiracle impedes the flow of water passing in through the mouth. A carefully programmed series of changes takes place; in some species in the space of just a couple of days; in others it may take as long as 10 days.

Externally the head becomes frog-like as a result of a spurt of growth of the head including the lower jaw. The larval mouthparts are shed, exposing the inside of the mouth and, within a day, the tadpole features have been lost and the creature assumes the appearance of a frog.

The tail rapidly shortens (in northern species, in water at 40°C, we have seen the entire process completed in twelve hours). It is accompanied by the accumulation of black (melanophore) pigment cells at the extremity.

PLATE 30 *Ian Morris with millions of tadpoles clumped together upon the Northern Territory flood plain. (G. Miles)*

PLATE 31 *A handful of tadpoles from Plate 30. (G. Miles)*

PLATE 33 Taudactylus eungellensis *from mountain torrents in the Eungella National Park in north central Queensland. (B. Stankovich-Janusch)*

PLATE 32 *Tadpole of a* Mixophyes *species changing into a frog. (M. Trenerry)*

This phenomenon may be explained by evidence that reabsorption of the tail is not accomplished by lymphocyte cells but by a form of cell death termed 'apoptosis' (Kerr, Harman and Searle, 1974). This is characterised by the absorption of particles of destroyed cells by adjacent cells. It would seem that these cells are unable to engulf melanophores.

The departure of the metamorphling from the water usually takes place before the tail stump is fully absorbed. The commencement of its existence on land represents the most vulnerable stage in its life as it faces numerous predators and has to find a spot where it can avoid dehydration.

CHAPTER EIGHT

LIFE IN ARID AREAS

BURROWING

I was tempted to title this chapter 'Life in the Deserts', which sounds more impressive. In reality the techniques employed by frogs to survive and breed in deserts are similar to those used in much less extreme environments where water is scarce or lacking for a portion of the year.

The most fundamental adaptation to periods of aridity is the capacity to burrow and, by this means, use the soil as an insulator. Frogs with this ability range throughout the entire Australian mainland and Tasmania too. Thus adaptations to the environmental extremes found in deserts, are simply an elab-

oration and perfection of an extremely common phenomenon. They occur, for example, in many species of frogs occupying the so-called wet/dry tropics of northern Australia where one-third of the year is extremely wet, and two-thirds extremely dry.

Approximately 35 per cent of the Australian frog fauna can burrow, whilst many more, to avoid dehydration, hide down cracks in the soil or beneath boulders or vegetation, so achieving a similar objective without actually excavating the soil (Figure 32).

That such a high percentage of Australian species can live in arid zones is probably a consequence of a long history of adaptations to drought. In their book on the Australian

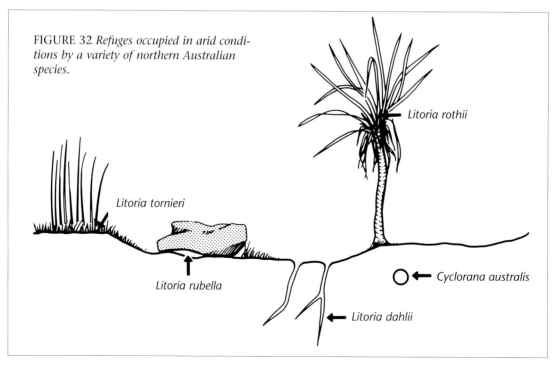

FIGURE 32 *Refuges occupied in arid conditions by a variety of northern Australian species.*

Litoria rothii

Litoria tornieri

Litoria rubella

Cyclorana australis

Litoria dahlii

TABLE SEVEN: Examples of burrowing frogs

BACKWARDS SLIDING BURROWERS	CIRCULAR BURROWERS	HEAD-FIRST BURROWERS
Cyclorana australis	*Neobatrachus* sp.	*Arenophryne rotunda*
Cyclorana longipes	*Notaden bennetti*	*Myobatrachus gouldii*
Limnodynastes dumerilii	*Notaden melanoscaphus*	
Limnodynastes ornatus	*Notaden nichollsi*	
Litoria alboguttata		

Sources: Sanders and Davies (1984), Tyler (1976), Tyler *et al.* (1980).

climate Linacre and Hobbs (1977) state, 'seasonal droughts are part of the way of life in Australia, but unseasonal droughts are less predictable and may be ruinous'. As an example of the unreliability they state that an area of 250 000 km^2 in Queensland, 'experienced its wettest year in 1950 and its driest year in 1951'. They go on to observe, 'An essential characteristic of Australian rain is fickleness'. Annual fluctuations are shown in Figure 6, page 20.

In some geographic areas of extreme aridity frogs may not survive. For example, when Cooper's Creek flows from Queensland to Lake Eyre in South Australia, it transports aquatic animals including frogs. If the flow is followed by a long period of drought in this, the driest region in the continent, the frogs will die. Perhaps five or ten years will elapse before a second flood transports more animals to Lake Eyre and fills the string of intermediate waterholes. Thus at this extreme the pattern will be one of repeated colonisations commonly interspersed with extinctions. In general, however, droughts are of shorter duration and burrowing provides the most common means of surviving them.

The species that burrow have a quite distinctive form: a broad head, bulbous body and short limbs. Almost all burrowers have a similar modification of one or two structures on the undersurface of the foot, beneath metatarsal bones, and termed the 'metatarsal tubercles' or 'scaphoid'. The modification consists of a downwards extension of the normally flattened tubercles to form a sharp cutting edge (Figure 33). They use these modified structures to scrape the soil out from beneath their bodies. But for it to be effective the soil must be moist or sandy. Burrowing through a solid surface can produce horrific injuries, as observed by Tyler, Davies and Walker (1985) who documented damage to frogs that had fallen into an empty swimming pool and attempted to burrow through the concrete floor.

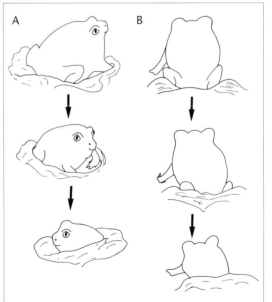

FIGURE 34 *Variation in burrowing techniques: A. Circular burrowing; B. Backwards sliding. See* Table 7 *for examples. After Sanders and Davies (1984).*

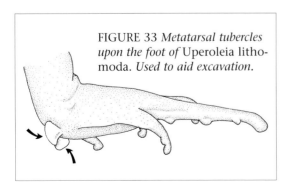

FIGURE 33 *Metatarsal tubercles upon the foot of* Uperoleia lithomoda. *Used to aid excavation.*

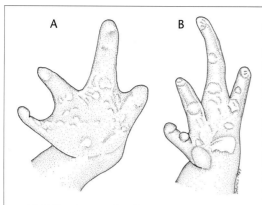

FIGURE 35 *Hands of: A.* Arenophryne rotunda *which uses them for burrowing and B.* Crinia deserticola *which does not burrow.*

Sanders and Davies (1984) found that there is nothing haphazard about burrowing but that there are two kinds of 'backwards burrowers', who differ from one another in the manner in which they excavate. Each follows a particular behavioural sequence of movements. One group they termed 'backwards sliding burrowers', which descend as they push backwards, so entering the soil at an angle. The second group they called 'circular burrowers', which are frogs that descend straight downwards (Figure 34). Examples are listed in Table 7.

There remains a minority group, of per-haps just two species, that use their hands and burrow head-first. The first to be documented was the sandhill frog *Arenophryne rotunda* whose burrowing behaviour is described in Chapter 11.

Headfirst burrowers have a small protective pad on the tip of the nose and a spade-like hand with short, broad fingers (Figure 35).

SKIN STRUCTURE

Almost all water loss occurs through the skin and, in fact, in most species water can evaporate from the surface of the skin as readily as from the surface of a pond. Thus a fundamental structural modification to life in dry areas would seem to require a waterproof skin. Curiously this attribute has not evolved in burrowing species; instead some of them develop an external and rather impervious cocoon derived from several layers of dead outer skin.

The outer surface of frog skin is an area where the formation of new cells (cytogenesis) and the death of older ones (cytomorphosis) is a rapid process. Cytomorphosis occurs progressively as the cells are pushed to the outside by new cells forming in deeper tissue beneath them (Figure 36). Periodically the outermost layer is shed as a complete sheet; a process known as 'sloughing' which can often be as frequent as once each week.

FIGURE 36 *Diagrammatic representation of the progressive stages in the formation and death of layers of cells in the skin, and the production of a cocoon. A. Normal skin: the youngest cells are those at the base, and the oldest those nearest to the surface; B. The outer layer has been sloughed off, whilst a new layer forms at the base; C. Following a further slough the cocoon is now two cells thick.*

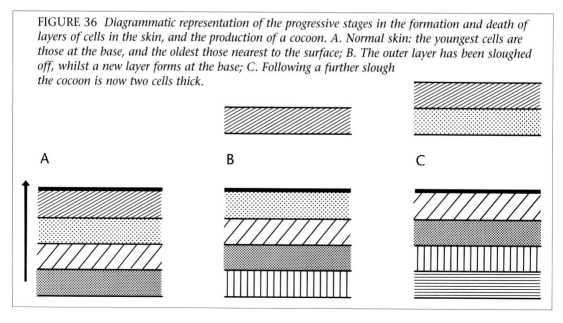

The removal of the outer coat is brought about by the frog itself. The frog undergoes slow but extreme contortions, whilst alternately inflating and deflating its lungs. As a result of these stresses the skin eventually splits on the flanks or behind the head. The frog then contorts and stretches until it is able to get hold of one edge of the split skin and stuff it into its mouth. Then with the aid of its hands it slowly pushes more and more into the mouth until the entire old skin is swallowed. It is not a pretty sight!

Cocoon formation in burrowing frogs is a modification of the sloughing process. The outer layer of dead cells separates from the cells beneath but is not shed.

Instead it hardens and remains as an outer cover like a shiny plastic bag (Plate 34). As the cocoon hardens it becomes much less pervious to water; in cocooned *Litoria alboguttata*, Seymour and Lee (1974) found that evaporative water loss was reduced by approximately seven-and-one-half times, compared with individuals lacking cocoons. Van Beurden (1984) found that in *Cyclorana platycephala* the presence of a cocoon delayed the loss of 10 per cent of body weight from eight hours in frogs lacking a cocoon to almost eight days.

LYMPHATIC SACS

A characteristic feature unique to frogs is the manner in which the skin is attached to muscles and other underlying tissues. Instead of being attached directly, the skin is separated by sheets of very thin, transparent connective tissues termed 'septa'. Thus the skin is loose and the septa create a series of pockets beneath it.

The function of the subcutaneous lymphatic sacs had been assumed to be sites for the storage of water. However Carter (1979) showed that the sacs are reduced in fossorial species (which one would have assumed most likely to benefit from such stores). They were most highly developed in aquatic species (the ones that would seem least likely to need to store water). Carter, therefore, proposed that the subcutaneous sacs are part of a system for the rapid excretion of incoming water. The sacs in a fossorial species are compared with those of a tree frog in Figure 37. The reduc-

PLATE 34 *Water-holding frog* Cyclorana longipes *inside its transparent cocoon. (P. Kempster)*

tion of the capacity of the sacs in fossorial frogs, in some areas of the body, by direct adherence between skin and body wall no doubt reflects the fact that there is no need for rapid excretion of water in burrowing species. But it also may mean that water loss is reduced when the skin is fused to underlying tissues.

REPRODUCTION

Survival in seasonally arid areas is such a feat in itself, that to manage to reproduce there is little short of remarkable. For a long time it has been considered that, in arid areas, the frogs are opportunistic breeders ready and able to breed immediately any rain falls. This assumption may be inaccurate. Van Beurden (1979) has shown that the best known of all arid-adapted species, the water-holding frog *Cyclorana platycephala*, does not respond to rains falling outside its assumed breeding season. The questions seem to be: how much rain, and when?

If a frog burrows to a depth of say 0.5 m in clay soil, dry times will convert that soft layer of clay to the consistency of a brick. It follows that some rains will scarcely penetrate the hardened clay, and that it requires a very heavy downfall to flood the ground for long enough for water to reach the entombed frog. This seems to be an excellent safety valve. For

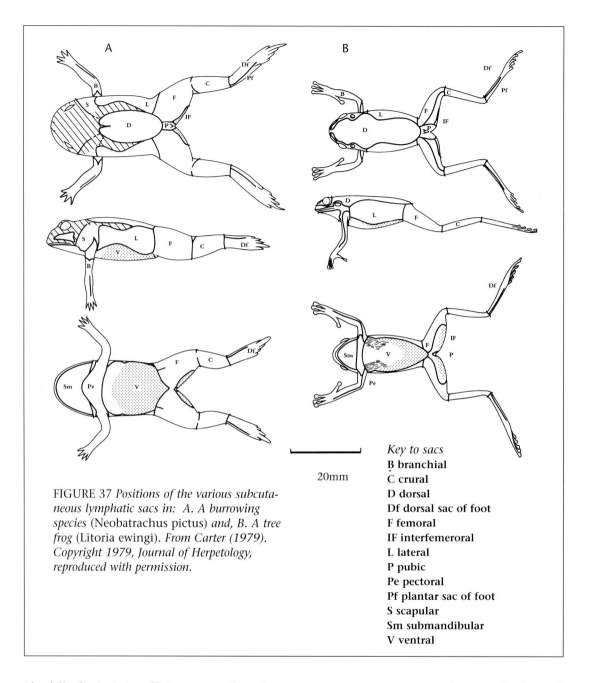

FIGURE 37 *Positions of the various subcutaneous lymphatic sacs in:* A. *A burrowing species* (Neobatrachus pictus) *and,* B. *A tree frog* (Litoria ewingi). *From Carter (1979). Copyright 1979, Journal of Herpetology, reproduced with permission.*

Key to sacs
B branchial
C crural
D dorsal
Df dorsal sac of foot
F femoral
IF interfemeroral
L lateral
P pubic
Pe pectoral
Pf plantar sac of foot
S scapular
Sm submandibular
V ventral

if a fall of rain is insufficient to reach and so release entombed frogs, it is equally unlikely to persist for long enough for spawn and then tadpoles to complete their development.

Developmental spans are influenced by temperature and, when rains fall in summer, water temperatures can be very high, and developmental rates accelerated. Spawn and tadpoles frequently are exposed to temperatures greater than 35°C. Those obtained by my colleagues and I in northern Australia are included in Table 8.

TABLE EIGHT: High temperatures experienced by spawn or tadpoles in northern Australia*

TEMPERATURE °C	SPECIES AND STAGE	REMARKS
45.0	*Litoria coplandi* tadpoles	Small rock pool above East Mitchell River Falls, WA. 15.2.79
43.0	*Cyclorana australis* tadpoles	Gravel pit, Newry Stn, NT. 15.2.86
42.5	unidentified tadpoles	Shallow pool, Bing Bong Stn, NT. 9.2.89
42.4	unidentified tadpoles	125 km S. of Katherine, NT. 13.2.83
42.2	*Litoria rubella* spawn	26 km N. of Daly Waters, NT. 16.12.80
41.9	unidentified tadpoles	Shallow pool, Bing Bong Stn, NT. 7.2.89
41.7	unidentified tadpoles	Shallow pool, Phillips Range, Kimberley Divn, WA. 3.2.85
41.2	*Cyclorana australis* tadpoles	Pool on blacksoil plain, Newry Stn, NT. 9.2.86
41.1	unidentified tadpoles	Dam, 250 km N.E. of Derby, WA. 1.2.85
41.0	unidentified tadpoles	Roadside pool, Delamere Rd, 113 km S. of Victoria Hwy, NT. 11.2.83
40.6	unidentified tadpoles	Roadside pool, Delamere Rd, 30 km S. of Victoria Hwy, NT. 11.2.83
40.6	unidentified tadpoles	Roadside pool, Delamere Rd, 113 km S. of Victoria Hwy, NT. 11.2.83
40.4	unidentified tadpoles	Roadside pool, 28 km N. of Halls Creek, WA. 6.2.82
40.2	hylid tadpoles	Buchanan Hwy, 117 km E. of Top Springs, NT. 11.1.83
39.6	*Litoria rothii* tadpoles *Litoria rubella* tadpoles *Litoria inermis* tadpoles *Cyclorana cryptotis* tadpoles *Uperoleia lithomoda* tadpoles	Inundated grassland adjacent Lake Argyle, WA. 11.2.87
39.5	*Litoria rothii* tadpoles *Litoria rubella* tadpoles *Litoria inermis* tadpoles *Cyclorana cryptotis* tadpoles *Uperoleia lithomoda* tadpoles	Inundated grassland adjacent Lake Argyle, WA. 7.2.87
39.5	*Litoria rothii* tadpoles *Litoria rubella* tadpoles	Inundated grassland adjacent Lake Argyle, WA. 9.2.87

*Unpublished data compiled by M. Davies, A.A. Martin, M.J. Tyler and G.F. Watson.

TEMPERATURE °C	SPECIES AND STAGE	REMARKS
	Litoria inermis tadpoles *Cyclorana cryptotis* tadpoles *Uperoleia lithomoda* tadpoles	
39.2	*Uperoleia inundata* spawn	Jabiru, NT. 11.1.81
39.0	*Litoria rothii* spawn	Victoria Hwy, 2.5 km E. of Victoria R., NT. 2.2.82
39.0	*Litoria rubella* spawn	Jabiru, NT. 12.1.81
38.8	*unidentified* tadpoles	Delamere Rd, 34.4 km S. of Victoria Hwy Junction, NT. 11.2.83
38.5	*unidentified* tadpoles	Gravel scrape 222 km N.E. of Derby, WA. 1.2.85
38.0	*Cyclorana longipes* tadpoles	Flooded paddock. Bing Bong Stn, NT. 10.2.89
37.9	*Cyclorana australis* tadpoles	Pool 117 km E. of Keep R. on Victoria Hwy, NT. 7.2.86
37.9	*Litoria rothii* tadpoles *Litoria rubella* tadpoles *Litoria inermis* tadpoles *Cyclorana cryptotis* tadpoles *Uperoleia lithomoda* tadpoles	Inundated grassland adjacent Lake Argyle, WA. 13.2.87
37.7	unidentified tadpoles	Small pool, King Leopold Range, WA. 4.2.85
37.5	*Cyclorana australis* tadpoles	Cattle trough. Broome, WA. 6.2.94
37.2	unidentified tadpoles	Delamere Rd, 113 km S. of Victoria Hwy Junction, NT. 11.2.83
37.2	unidentified tadpoles	41 km S. of Top Spring, NT. 12.2. 83
36.9	hylid tadpoles	Victoria Hwy, 2.5 km E. of Victoria River, NT. 2.2.82
36.5	unidentified tadpoles	Large pool, 89 km N.E. of Derby, WA. 4.2.85
36.1	*Litoria* coplandi	Little Adcock River, Gibb River Rd, WA . I.2.85
36.1	*Litoria rothii* tadpoles *Litoria rubella* tadpoles *Litoria inermis* tadpoles *Cyclorana cryptotis* tadpoles *Uperoleia lithomoda* tadpoles	Inundated grassland adjacent Lake Argyle, WA. 10.2.87
36.0	*Cyclorana sp.* tadpoles	Buchanan Hwy, 117 km E. of Top Springs, NT. 12.2.83
35.3	*Litoria rubella* tadpoles	Shallow dam at Great Sandy Desert, WA. 30.1.85
35.2	*Cyclorana cultripes* tadpoles	Large dam at Alroy Downs, NT. 15. 12.81

PLATE 35 Bufo marinus *tadpoles forming an aggregation in a shallow pool 82 km west of Hughendon, Queensland. (M.J. Tyler)*

BASKING

Survival in seasonally arid country appears to be achieved by a combination of behavioural and structural ploys that minimise water loss and avoid exposure to extreme heat.

Tyler showed (1976b) a photo taken by G. Gow of *Notaden melanoscaphus* adopting a peculiar pose in which its back legs are stretched out beyond the body; the only possible interpretation of the unusual behaviour was that the frog was maximising the body surface exposed to the sun, so intending to elevate body temperature.

Within my experience, *Cyclorana australis* basks frequently in the wet season. Instead of burrowing to avoid the heat of the day, the frogs sit in an exposed position very close to the edge of pools. They give the impression of being asleep and, provided that you walk slowly and carefully to minimise noise and vibration, it is possible to walk up to them and simply pick them up. Body temperatures and temperatures of the adjacent environment are documented in Table 9.

TABLE NINE: Temperature (C°) of basking *Cyclorana australis* and the associated enviroment*

CLOACA	BACK	SOIL	AIR	LOCALITY
24.5	—	29 . 2	31. 2	9 km west of Tennant Creek, NT. 17.12.80
—	—	36.2	37.0	Banka Banka Station, NT. 17.12.80
33. 4	33. 0	32. 0	—	Jabiru, NT. 30.11.77
34. 2	34. 4	34. 0	—	Jabiru, NT. 30.11.77

* M. Davies, A.A. Martin and M.J. Tyler unpublished data.

CHAPTER NINE

LIFE ABOVE THE GROUND

FINGER AND TOE DISCS

Just how frogs in trees or upon rock faces stop themselves from falling off is a topic that has fascinated biologists for years. What climbing frogs possess, that are not shared by other frogs, are enlarged, flattened discs on the tips of their fingers and toes. It follows that there has been a great deal of study, and more speculation, about how these structures aid life above the ground. Obviously they increase the surface area, but do they stick? Is there suction involved? Resolution of these issues has occurred only recently.

Green (1979) used a scanning electron microscope to examine cells on the surface of the discs, revealing a pattern of interlocking, polygonal cells with narrow gaps between them. It is apparent that similar cells probably occur on the disc of all arboreal frogs (for example, see that of *Litoria rothii* in Figure 38).

Emerson and Diehl (1980) argued that two separate mechanisms are involved in the way in which discs aided climbing. The first was one of an interlocking of the surface cells against tiny irregularities upon rough surfaces. The second occurred on smooth surfaces and was said to be 'capillarity', requiring close contact of two moist surfaces. It was thought that the canal-like network between the polygonal cells would drain away excess mucus, and aid in establishing a thin, even layer for adhesion to occur.

Green (1981) noted that the efficiency of adhesion was related to the surface area of the discs, and not to the number of cells, and argued that surface tension is the primary means of adhesion. As shown by Tyler (1976b, Plate 37) the loose skin of the chest

and abdomen increase the area of contact between a frog and a smooth, vertical surface. Green points out that the discs alone are sufficient, but I suspect that this is so only for short periods, and that the ventral skin has a role to play when the frog is at rest for any length of time.

Most arboreal frogs are slightly built, with slender bodies and limbs (Plate 15). Exceptions, such as *Litoria caerulea* and *L. splendida*, are much more robust and the latter has huge glands on the dorsal surface of the head (Figure 39). Green (1979) has shown that the enlarged areas beneath the finger and toe joints (subarticular tubercles) of *L. caerulea* have a similar surface structure to the discs, and that they unquestionably function as accessory discs.

It would be a simple matter to believe that if some tree frogs with discs on the tips of their fingers and toes, can rest on vertical surfaces, it can also be assumed to be so for other similarly equipped species. This generalisation is not correct. Observations on numerous species in my laboratory show that, whereas some will rest for days or even months in a vertical position on the wall of an aquarium, some never do. It seems that there is a critical body mass, above which the efficiency of the body surfaces devoted to attachment are inadequate to maintain purchase. Slender species such as *L. chloris* and *L. gracilenta* habitually attach to a vertical surface (Plate 39). But in *L. caerulea* and *L. splendida*, juveniles can attach but the heavy, bulky adults select horizontal sites for resting.

In all tree frogs, the discs are not efficient if they are wet; a frog climbing out of the water in an aquarium slips and slides down the glass wall.

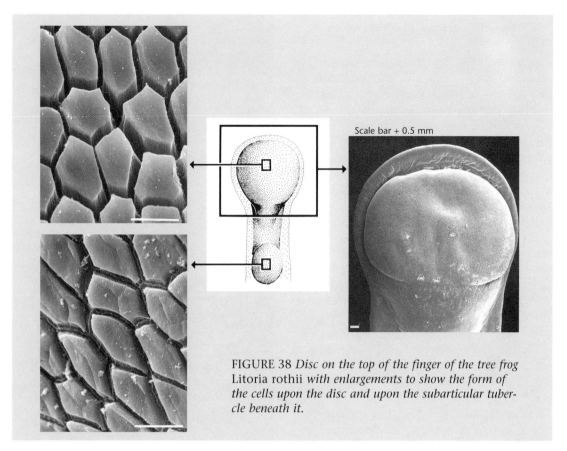

Scale bar + 0.5 mm

FIGURE 38 *Disc on the top of the finger of the tree frog* Litoria rothii *with enlargements to show the form of the cells upon the disc and upon the subarticular tubercle beneath it.*

A characteristic of the Australian species of *Litoria* and *Nyctimystes,* and shared by other members of the family Hylidae overseas, is the presence of an additional skeletal element between the terminal and penultimate phalangeal bones of each finger and toe. These structures were once called intercalary cartilages but, with the discovery that in many species they are boney, they are now termed 'intercalary structures' or 'intercalary elements'.

Duellman and Trueb (1986) state that intercalary structures of tree frogs enable the entire surface of the disc to come into contact with the substrate. Curiously amongst Australian species the largest intercalary structure occurs in *Litoria nasuta* (Plate 36), a species with greatly reduced discs, which does not climb and in fact lives upon the ground. The intercalary structure in *L. nasuta* is an elongate rod of bone. In contrast most other species have small discs of cartilage or bone (Figure 40).

If you or I tried to jump in a frog fashion, landing simultaneously upon hands and feet, we would no doubt break fingers as a result of their sudden compression against a nonpliant surface. It is tempting to view the function of intercalary structures in frogs as a device for absorbing that shock. But the reality is that the biggest frogs in the world that leap around (mostly *Rana* species) do not have intercalary structures. Why then should they be most highly developed in a relatively small, terrestrial species? A solution has yet to be proposed.

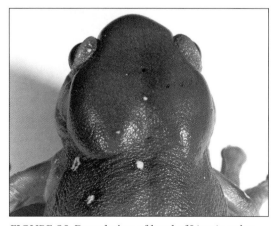

FIGURE 39 *Dorsal view of head of* Litoria splendida *showing the vast glands upon its surface.*

TEMPERATURE TOLERANCES

A good deal of effort has been devoted to the relationship of frogs with their surroundings, and the way in which they obtain their physical needs. In particular, emphasis has centred upon avoidance of high environmental temperatures, and the manner of maintaining water balance.

Numerous contributors have undertaken experiments designed to determine the upper and lower temperatures at which frogs can survive. By far the most extensive survey of Australian frogs is that of Brattstrom (1970) who examined the responses of 42 species. His technique involved maintaining frogs at any of several fixed temperatures for up to 100 hours, and comparing the survival of species transferred from high to low temperatures, and from low to high. During the experiment they were deprived of food and maintained at a constant level of illumination. Prior to use the specimens were kept at a fixed acclimation temperature.

Brattstrom concluded that cool-adapted species are predominantly southern in distribution and warm-adapted predominantly northern, whilst species exhibiting wide ranges of tolerance to temperatures have the widest geographic range.

There is perhaps little that is surprising about the above conclusions, but the experimental design described bears little resemblance to conditions experienced in the field, where temperature and light fluctuate, and where feeding can take place each night.

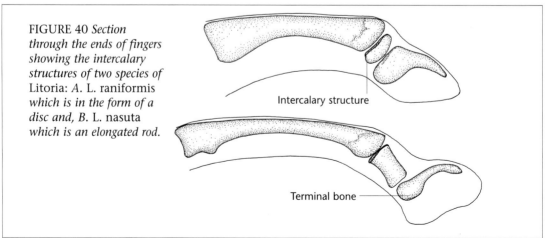

FIGURE 40 *Section through the ends of fingers showing the intercalary structures of two species of* Litoria: *A.* L. raniformis *which is in the form of a disc and, B.* L. nasuta *which is an elongated rod.*

Intercalary structure

Terminal bone

PLATE 36 Litoria nasuta. *(B. Stankovich-Janusch)*

There seem to be very little published data on diurnal body temperatures of tree frogs in their natural environment. The likelihood of finding specimens and having the opportunity to record body temperature is remote. An exception involves *Litoria rothii*, a northern Australian species (from which Brattstrom, 1970, recorded in the laboratory a tolerance to temperatures of up to about 39°C). In the jackeroos' washhouse on Newry Station, Northern Territory, at midday on 8 February 1986, Graeme Watson and I spotted two *L. rothii* resting upon the upper surface of a wooden beam supporting the corrugated iron roof, and perhaps 2 cm from the metal.

The body temperatures of the frogs (measured by inserting the tip of a slender-bulbed thermometer into the cloaca), were 33.1 and 33.4°C respectively. Air temperature within the washhouse was 38°C and the relative humidity 46 per cent. It was not possible to measure temperature upon the ledge where the frogs were located; but the roof above was far too hot to touch and a temperature of 40–45°C would seem likely upon the ledge beneath.

How then could the *L. rothii* maintain a body temperature perhaps as much as 10°C

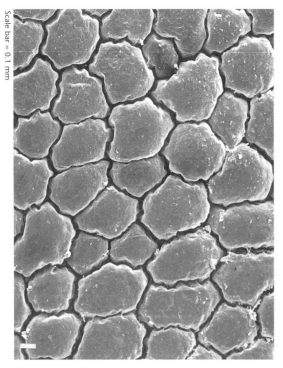

Scale bar = 0.1 mm

FIGURE 41 *Scanning electron microscope view of the granular skin upon the ventral surface of the body of* Litoria peronii.

below that of the environment? It seems too glib to trot out the usual statement that a frog's body temperature largely is determined by the environmental temperature, and that it is only by behavioural ploys that it is able to reduce or elevate it.

Elevation of body temperature is simple; all a frog has to do is to get out there and expose its body to the sun. But it is a far more difficult problem to conceive of ways in which a frog can reduce its temperature, particularly in the example cited where the surrounding temperature would have been fairly uniform. Certainly the skin of *L. rothii* in hot conditions fades to a pale putty colour in direct sunlight, minimising the heat that they absorb from the surroundings but, without any internal control for an animal in a uniformly hot environment, the only cooling mechanism would seem to be an evaporative one. For that to be effective, the release to the surface of a significant quantity of water is required.

WATER BALANCE

For a frog in the artificial environs of a washhouse, restoration of water loss through evap-

oration would be a simple matter, but in the open there may not be an opportunity.

Some tree frogs have been shown to have a waterproof skin. Withers, Hillman and Drewes (1984) studied water loss in species from several continents, and included *L. caerulea* and *L. gracilenta* in their sample. They found that the dorsal skin of *L. gracilenta* was waterproof, whereas that of *L. caerulea* was not. Perhaps the most obvious behavioural difference between these species is the pose that they adopt at the onset of dry conditions. Like the *L. xanthomera* in Plate 39, *L. gracilenta* manages to get much closer to the substrate, tucking its fingers beneath the throat, and its toes beneath its legs. This action both reduces the surface area of the body exposed, and protects the thinnest structures of the body that presumably would be prone to damage by dehydration. The frog really looks as though it is settling down in preparation for a long wait.

The bulkier form of *L. caerulea* prevents it from adopting a similar pose. However after a period, the skin of both species becomes shiny, glistening like enamel, which might be due to the secretion of water-resistant lipids. Thus the findings of Withers *et al.* (1984) that *L. caerulea* is not waterproof is surprising: perhaps the pose is as significant as the nature of the skin.

If waterproofing is the exception rather than the rule amongst Australian tree frogs, good candidates would seem to be *L. chloris* and the closely related *L. xanthomera* because of their habit of remaining in fixed positions for long periods of time.

Frogs do not drink and, instead, obtain almost all of their water requirements by absorbing it through the ventral skin. The only possible supplement would be the water content of various food items.

Water intake through the skin is a simple process whereby the spaces between the cells of the body develop an increasingly negative pressure as water is lost (Stromme, Maggert and Scholander, 1969). This negative pressure constitutes a force capable of pulling water from the skin into the body. From the skin, water enters a subcutaneous lymphatic sac and is transferred into a pulsating lymph heart which drives the fluid into the bloodstream via narrow lymphatic vessels. The internal transit system is extremely efficient: water passing into the dorsal sac enters the blood stream in less than 15 seconds.

Many species of frogs that have been exposed to artifical drying conditions in the laboratory and then provided with a moist surface, assume a characteristic pose. They flatten themselves and shift the foot from its customary position beneath the tibia to a position beside it. What they are doing is increasing the surface of the hindlimbs exposed to the moist substrate, so increasing the surface area from which water can be absorbed.

This 'water absorption response' as it is called, was first recorded by Stille (1958) in nine species and four genera of North American frogs, and he predicted that it may prove to be widespread amongst frogs. Johnson (1969) confirmed that it took place in the five genera and 15 species of Australian hylids and leptodactylids that he examined.

Clearly the response is not confined to tree frogs, but it is amongst this group that modifications of the skin to aid the water absorption response are most highly refined.

What tree frogs have developed is an increase of the surface area of the ventral skin in contact with the moist substrate, no doubt improving the efficiency of water absorption. The skin of the ventral surface, including that of the hindlimbs, is granular, consisting of regularly-shaped, protuberant granules with narrow spaces between them (Figure 41). The spaces may aid water dispersal across the skin by means of a capillary pull.

Granular skin is most highly developed upon those parts of the skin that come into contact with the moist substrate during the water absorption response.

PLATE 37 *Dainty green tree frog,* Litoria gracilenta, *with distended vocal sac. (M. Trenerry)*

PLATE 38 *The smallest Australian tree frog* Litoria microbelos. *(M. Trenerry)*

PLATE 39 Litoria xanthomera *at rest upon the wall of an aquarium. (B. Stankovich-Janusch)*

CHAPTER TEN

THE CANE TOAD

In the 50 years that have passed since its introduction into Queensland to control insect pests of sugar cane, the cane toad has been the subject of considerable debate and innuendo. But over the past decade more data have been gathered than in the previous 40 years, and it is at last possible to sift a few of the facts from fiction.

Overall, there is need for action to at least reduce the number of toads, if not to eliminate them altogether, and control action is now being examined as a matter of extreme urgency.

The fundamental omission from the data considered by the authorities when the introduction of the toad into Australia was mooted, was attention to the size of Australia. It was the success of the introduction of the cane toad to Puerto Rico, up until 1932, that catalysed events and led to its introduction into Australia. But Puerto Rico is an island only marginally bigger than Kangaroo Island, South Australia. The toad population had only limited land to occupy. In contrast, the canefields of Queensland were not marooned or isolated in any way and, as the cane toad numbers increased, many of them dispersed into the surrounding countryside.

CANE TOAD IMMIGRANTS

The cane toad is native to an area extending from the northern portion of South America, throughout Central America to the southern part of Texas in the USA. This is an extensive geographic range and includes both tropical and seasonally arid areas. For a species of animal to exist in such a remarkable range of environments it must be extremely adaptable, and tolerant to a variety of physical conditions. With the wisdom of hindsight it should have been predicted that the cane toad was likely to become established almost anywhere. But no-one could have predicted how rapidly or how effectively.

Quite a few species of frogs have been taken from one country and deliberately released in another. A good review of the introductions into Australia of all sorts of animals is presented by Frith (1979). At first hearing, the concept of introductions without any real reason may seem fairly pointless, and there is little evidence to cause anyone to change that first impression. Often early settlers just set out to remedy what they viewed as deficiencies in faunas. Some animals were introduced for hunting, some escaped from zoos. There were exports from Australia too. Thus some Australian tree frogs were released in southern England and thrived for a few years but during a severe winter they died. Other species of Australian frogs were introduced into New Zealand, some perished but others survived there. Some of these introductions were of little ecological consequence, some were disasters; one of the disasters was the introduction of the cane toad into at least 15 countries.

Just who was the first person to become engaged in the cane toad trade is uncertain. But prior to 1844 cane toads had been shipped from French Guiana to Martinique, then others from the newly established Martinique population were liberated in Barbados, and later still toads went from Barbados to Jamaica. The history of the introductions is sketchy but it is known that they were attempts at biological control. The huge toads were brought in to control pests ranging from insects to rodents.

In Europe and North America, gardeners

have always tended to regard toads as useful animals to be protected, because they are aware that they will eat insects, snails and other creatures that are harmful to cultivated plants. The introduction of the cane toad into parts of the West Indies was an application of this knowledge.

But it wasn't until the 1920s and 1930s that the cane toad was dispersed on a large scale. The chain of events started in 1920 when Dr D.W. May, Agronomist-in-Charge of the Porto Rico Agricultural Experimental Station imported and released in the station grounds 12 cane toads from Trinidad.[1] By 1924 Stuart Danforth of Cornell University in New York reported that 'they had become quite common within a radius of four miles', but in the same year the stock was supplemented by a further shipment of 40 specimens, this time from Barbados. The object of these introductions was to control the beetle *Phyllophaga vandinei* which had become a serious pest of crops such as banana, coconut, breadfruit and even sugar cane. Whereas the beetles stripped foliage of plants, their larvae fed on the roots of the sugar cane.

The sugar cane crop was vital to Puerto Rico's economy and there were no known control measures against a beetle infestation that was becoming extremely serious. In fact at that time the beetles were being collected by hand and the sugar producers paid a bounty on each pest beetle they brought in. Smyth (1916) cites data from the Guanica Central plantation: 'During six months from 27 November 1913 to 14 May 1914, the Central paid for the collection of 1 662 000 *Phyllophaga*, $1876.73'. (If the number of beetles seems astronomical, the idea of 1.1291 cents per beetle seems an odd rate to offer!)

Faced with a potential crisis situation the cane toad was viewed as the answer to plantation owners' prayers. Here was an animal that bred like rabbits, required no maintenance and, without question, would devour every insect it encountered. Not only was it more efficient than human collectors but it was free! No more bounty payments and, hopefully, no more plagues of pests!

It all seemed too good to be true. Puerto Rico had found a solution to a serious problem common to sugar cane producers throughout the Pacific region. It was only a matter of time before other producing countries learned of the technique and tried it out for themselves.

In March 1932 the International Sugar Cane Technologists held their fourth annual meeting and, significantly, Puerto Rico was the host country. By that time the cane toads there had achieved more than could have been hoped for. The frenchi beetle was no longer a serious pest. On their tours of the cane fields the conference visitors could see for themselves that Puerto Rico had achieved what other sugar cane producers had not imagined possible.

The most relevant report presented at the conference was one by Raquel Dexter who demonstrated that on a percentage basis (in terms of the frequency of food items) the stomachs of *Bufo marinus* contained 41 per cent *Phyllophaga* and a second pest *Diaprepes*. In areas where there was plenty of food for the *Phyllophaga* beetles, Dexter calculated that on average each toad ate 12.5 beetles per night and, applying this to Smyth's data, pointed out that 1000 toads would have eaten 1 662 000 beetles (the number cited by Smyth) in 133 nights. These were dramatic words and overnight the cane toad became almost sacred! The journal *Nature*, not renowned for exaggeration, headed an article published in November 1934: 'Toads save sugar crop'. Not surprisingly there was an immediate move to import cane toads into many countries (summarised by Easteal, 1981).

By modern standards the introduction of the cane toads would be criticised for not having taken into account the effects that this animal might have on native animals other than the various 'target' species of insects it was planned to control. However any criticism has to take into account the massive problems faced by the agriculturalists, and particularly the entomologists specialising in insect control. Sugar cane production was being endangered by very serious insect pests, the entomologists lacked insecti-

---

[1]Easteal (1981) quotes the source of the material to be Barbados.

cide sprays or other weapons, and so the collection of insect pests by hand was obviously a last resort, and an expensive exercise of dubious value at that!

Today toads or any other animal harnessed to perform biological control would only be released following exhaustive tests proving that the introduced animals would not have deleterious effects upon the native fauna. But the fact that these checks are required now is simply because lessons have been learned from unfortunate experiences in which native fauna has suffered. Our predecessors didn't have the benefit of the experience that has produced the caution.

As shown in Table 10, Hawaii was the first country to act upon the Puerto Rican experience. Mr C.E. Pemberton of the Hawaiian Sugar Planters' Association visited Puerto Rico in March and April 1932 and collected 154 toads. Oliver (1949) writes that Pemberton 'packed them in his suitcase' but Pemberton (1933) states that he shipped them in four separate wooden crates packed with moist wood shavings! Three of the crates went by rail and streamer and the fourth by airplane, and they took 13 or 19 days to reach their destination. Pemberton (1933) states 149 toads arrived in good condition; Oliver (1949) states 148. Perhaps these may seem trifling points of detail, but so much rubbish and innuendo is associated with the toad that it is worth noting how even the most fundamental facts become distorted.

The toad rapidly became established in Hawaii. Within 17 months of the liberation of the original stock the toads had multiplied to such an extent that 105 517 of their offspring had been redistributed elsewhere on the islands (Pemberton, 1934). And because Hawaii was so much closer to the nations of the south-west Pacific margin than Puerto

TABLE TEN: Sequence of major introductions of the cane toad

COUNTRY	SOURCE OF TOADS	DATE OF INTRODUCTION
Martinique	French Guiana	prior 1844
Barbados	Martinique	" 1844
Jamaica	Barbados	1844
Bermuda	British Guiana	? 1885
Puerto Rico	Trinidad	1920
" "	Jamaica	1924
Hawaii	Puerto Rico	1932
Philippines	Hawaii	1934
Australia	"	1935
Formosa	"	1935
Fiji	"	1936
New Guinea	Australia	? 1937
New Britain	"	1937
New Ireland	New Britain	1939
Ellice Islands	Fiji	1939
Solomon Islands	"	1940
Marianas Islands	?	prior 1944
Cuba	Puerto Rico	1946
Florida, USA	?	prior 1955
Admiralty Islands	?	?
Samoa	?	?
Palau	?	?
Thailand*	Australia	1976

* An accidental escape from a private zoo. Whether the cane toad became established is not known.

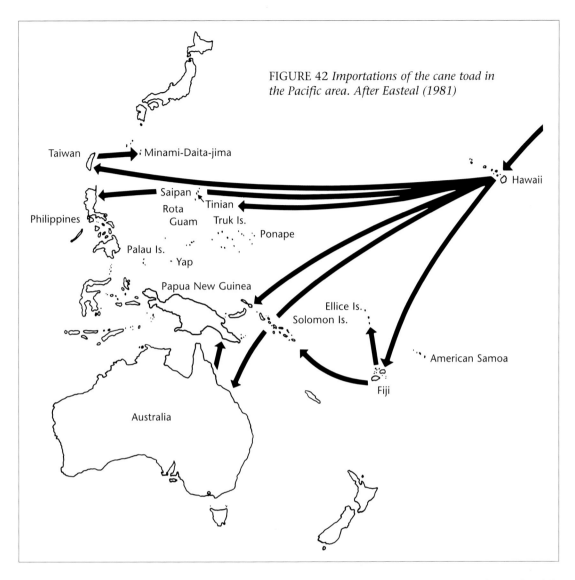

FIGURE 42 *Importations of the cane toad in the Pacific area. After Easteal (1981)*

Rico, it became a supplier for their initial shipments as well (Figure 42).

NATURAL HISTORY

The cane toad is a good example of what is called an opportunistic breeder. Whereas some species have a regular breeding season, and breed only within perhaps that two to three month period, others will breed whenever the opportunity presents itself. However it is important to recognise that this doesn't involve every individual. In New Guinea, Zug, Lindgren and Pippett (1975) found that in any one month 10 per cent of the females are ready to breed, but more males seemed to be in breeding condition from October through to January.

In a tropical environment where the difference in rainfall between seasons may not be as great as in temperate areas, a species with individuals breeding at any time will end up with occasions when it is the only species breeding on a particular day. If the food resource is limited, then the advantage to the species is huge, particularly when it lays as many eggs as the cane toad.

The breeding sites vary from ponds and dams to slow-flowing shallow streams and drainage ditches. Males congregate there, and their enthusiastic chorus of mating calls attract the females to the breeding site.

The eggs emerging are laid in long strands of jelly. In contact with water the jelly strands swell and form continuous, spotted streamers up to 20 m long which become entangled in aquatic vegetation (Plate 40). (Dexter (1932) reported the spawn of the cane toad to be in the form of a foam nest. She was clearly in error and must have been looking at the spawn of a native Puerto Rican species!) Each egg is surrounded by its own membranes and shares with other eggs the outer, cylindrical coat. Eggs measure 1.7–2.0 mm diameter and are black on the upper half. Females seem to lay 8000–25 000 eggs at a time. One popular article claimed 750 000!

The rate of development of the egg into a tadpole is determined almost entirely by the water temperature. Most reports in the literature suggest that the tadpoles generally hatch within 48 hours. Floyd (in Niven and Stewart, 1982) reports hatching to take place at about 25 hours at 34°C and to be delayed at lower temperatures, extending to 155 hours at 14°C. The tadpoles are small and black and have rather short tails (in comparison with other species). After a few days of frantic feeding the tadpoles form vast shoals involving thousands of individuals. Development proceeds rapidly and the tadpoles turn into tiny frogs after periods reported to range from 12 to 60 days. About one month seems typical for most ponds.

With the species name implying the ability at some stage to live in the sea, one would expect the tadpoles to perhaps be tolerant of sea water. It was reported to occur on beaches by Covacevich and Archer (1975) including Rainbow Beach. I found tadpoles and baby toads in running water on Rainbow Beach in November 1976 (Plate 41). I took a sample of the water there and later analysis showed it was fresh water, presumably draining onto the beach from the adjacent coastal area.

Ely (1944) did some experiments to test the tolerance of tadpoles to sea water and found that they are unable to complete development at anything more concentrated than 15 per cent sea water.

When the tiny baby toads (often only 5 mm long) emerge from their ponds or streams to start life on land they are especially vulnerable. Zug and Zug (1979) put it this way:

The likelihood of death for a marine toad is probably greatest from metamorphosis to subadult-hood. During this period, the toad is a juicy little morsel that has lost the toxicity of its larval stage and has not yet gained the protection of large body size and well-developed parotoid glands.

If it is their tiny size that makes the baby toads vulnerable to predation and dehydration, there is a great advantage for those individuals that grow rapidly. Zug and Zug's data show that growth generally is rapid for several weeks but that it then slows down a little until the animals become sexually mature, when body growth rates fall dramatically.

Given evidence of millions of baby toads emerging from the water the survival rate must be very low indeed. Zug and Zug (1979) estimate that out of any batch of eggs a maximum of only 0.5 per cent of the individuals will survive to attain sexual maturity.

Most adult toads encountered are within the size range of 90–150 mm, and large size is commonly associated with low population densities. There is always interest in knowing which is the largest individual and at present that distinction is held by a toad in the Queensland Museum (Figure 43). It measured 24.1 cm from the tip of its snout to the end of its body, and was 16.5 cm wide, and weighed 1.36 kg.

Just how old the really big toads are is hard to estimate. There is a record for this species of 16 years which was one of nine individuals maintained in a captive colony specifically to test how long they lived. The other eight survived for periods of eight-and-a-half to 14 years (Pemberton, 1949). There was no suggestion in Pemberton's report that the last survivor of his toads was particularly large. In Europe, toads which have distinct periods of seasonal growth, exhibit growth rings in sections of long bones and permit age calculations to be made with apparent accuracy. Similar studies have now been undertaken on *Bufo marinus* in Australia (Freeland, in manuscript).

Because the cane toad is expanding its range at such a substantial rate, there is considerable interest in knowing where it will become established and what parts of the continent are likely to remain toad-free zones. In terms of physical barriers the toads seem unlikely to penetrate forest. They tend to prefer open, cleared areas and are particularly well adapted to living near human habitation. But at a broader level of analysis much hinges upon the temperatures that they can tolerate, and their ability to withstand water loss. Stuart (1951) says the critical thermal maximum (the maximum body temperature that it can attain and survive at) is 42°C and the critical thermal minimum is 15°C. Krakauer (1970) estimated 40°C and 10°C respectively. Particularly significant, he found the critical thermal maximum of tadpoles to be 42.5°C. This upper limit is only slightly less than the maximum water temperature in which my colleagues and I have found tadpoles or spawn in the north of Australia in the wet season (Table 8). Obviously the tadpoles of the cane toad seem almost as well adapted to surviving in the hot temporary waters found in northern Australia as the native species.

The temperature tolerances of individual toads can be modified (that is either raised or lowered a few degrees) by getting them used to living at higher or lower temperatures for a few days before exposing them to the experimental temperatures that test the limits of their tolerance.

Tolerance of water loss is another important factor that will influence the ability of toads to colonise new areas. Whereas in moist, tropical conditions they can replace any water lost by taking it up through their skin at almost any time, survival in areas where they lack that chance will in part hinge upon the amount of water loss they can tolerate. Krakauer (1970) showed that they can tolerate up to 52.6 per cent reduction in body weight. He went on to demonstrate that tolerance of desiccation was an expression of the adaptation of species to living a terrestrial life —the aquatic species having a lower tolerance than those of the deserts.

THE AUSTRALIAN EXPERIENCE

In 1935 the decision was taken to introduce the cane toad into Australia to control pests of sugar cane. Mr R.W. Mungomery of the Sugar Experimental Station at Meringa in Queensland travelled to Hawaii, and on 1 June collected 102 toads. The consignment was despatched from Honolulu on 3 June and it reached Gordonvale, Queensland on 22 June. Only one of the toads failed to survive the journey.

The toads were taken to Meringa and the

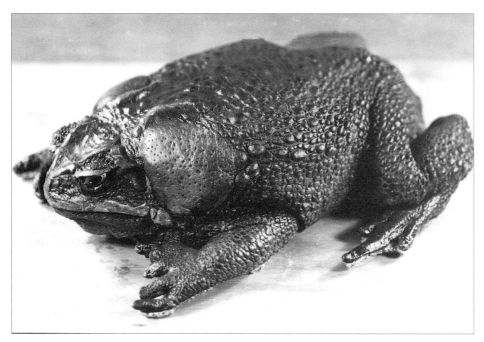

FIGURE 43 *Preserved gigantic cane toad now in the collection of the Queensland museum. It had a body length of 24.1 cm.*

following month during unseasonably warm weather the toads bred freely in captivity producing many thousands of offspring. As a result no less than 3400 young toads were taken and liberated in the areas around Cairns, Gordonvale and Innisfail.

However in August 1935 the Commonwealth government heeded expressions of concern from conservationists and banned the further liberation of the species. The information was conveyed to the Queensland Minister for Agriculture (Mr F.W. Bulcock) by the Federal Director-General of Health and Quarantine (Dr J.H. Cumpston) who advised Bulcock that there were fears that the toad would do harm by eating insects of economic value.

Understandably the news of the ban angered the canegrowers. In their journal *The Canegrower* of 27 November 1935 the feelings were expressed:

Mr Bulcock said that it was his intention to take the matter up with the Federal Government as soon as possible.

'The toads,' he said, were, in his opinion, 'the best economic contribution to the sugar industry for many years.' Very careful investigations were made about it, and an officer of the department had spent some months in Hawaii studying its habits for the express purpose of discovering whether the toad exhibited any of the tendencies Dr Cumpston attributed to it. That officer found no such tendencies. Nor did the Hawaiian expert, Mr J. Pemberton, who visited Australia with the sugar cane technologists a few months ago.

'Moreover,' said Mr Bulcock, 'since the toad had been liberated at Meringa a close watch had been kept on it, and if the toad had disclosed any of these undesirable characteristics' he would have been the first to hear about it.

'The fact is that the toad will not behave, nor will he [sic] behave, as Dr Cumpston alleges,' said Mr Bulcock. 'We have no concrete evidence that he becomes a menace to economically valuable insect life. We have definite and concrete evidence that the toads are valuable to the industry and will not be a menace.'

'Dr Cumpston cannot possibly have any evidence to support his contentions,' continued Mr Bulcock. 'He is apparently merely playing safe, at the expense of the unfortunate canegrowers in

grub-infested areas.'

Immediate representations were made to the Federal government seeking a lifting of the ban, and a temporary compromise was reached in September 1935. Because the toads at Meringa were continuing to breed, permission was granted to continue to liberate in the original areas of release, but not to permit releases in new areas. The result of this decision was a huge overstocking. Addressing a sugar industry conference reported on 9 April 1936, Dr Kerr stated:

. . . we have been obliged to liberate all young toads in the districts from Cairns to Tully. We have had hundreds of thousands of eggs liberated in ponds and streams in those districts.

By September 1936 the ban on liberation of the cane toad was lifted completely and toads were released in many new areas. Sabath, Boughton and Easteal (1981) undertook a detailed study of the way in which the cane toad colonised new areas following its release in the cane fields. They found that in the 40 year period 1935–1974 the population expanded at an exponential rate. They report that the species now is found in 584 000 km^2 which is equivalent to 33.8 per cent of the surface area of Queensland, and that the average annual growth rate of the geographic area is 8.1 per cent. Figure 44 reproduces an illustration prepared by Sabath and his colleagues plotting the expanding range in Queensland up to 1974, and the change that occurred through to 1980 (Easteal *et al.*, 1985).

There are other recent reviews of the distribution of the cane toad in Australia. The first was by Covacevich and Archer (1975) and the other by Van Beurden and Grigg (1980). The latter undertook two surveys of its distribution in north-eastern New South Wales. The first survey was conducted in 1974–1975 and the second in 1978. Previously it had been assumed that the New South Wales population was simply a southern extension of the Queensland boundary of the species. However they showed that the populations were separated by a gap of 25 km, and they calculated that at rates of spread

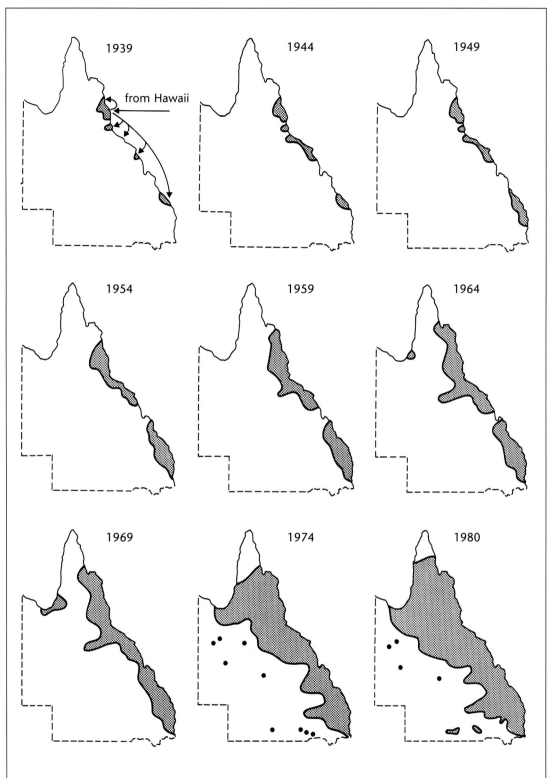

FIGURE 44 *Progressive expansion of the geographic range of the cane toad in Queensland. After* Sabath *et al.* (1981) and Easteal *et al.* (1985)

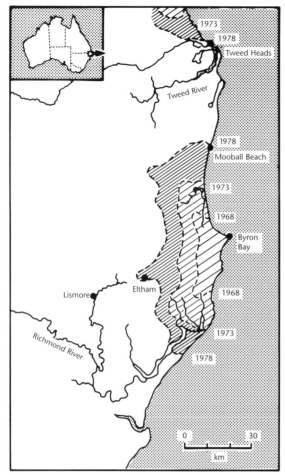

FIGURE 45 *Progressive expansion of the geographic range of the cane toad in New South Wales. After Van Beurden and Grigg (1980)*

of 1.07–3.0 km per year the populations would become united within four to twelve years (Figure 45). They commented, 'further dispersal southward appears certain'.

Expansion of the range of the cane toad by natural dispersal continues at an alarming rate. In 1983 it was discovered that the species had crossed the country at the foot of the Gulf of Carpentaria and had penetrated the eastern boundary of the Northern Territory (Figure 46). It is now well entrenched there, advancing at about 40 km per year, and it has been estimated that at its current rate of colonisation it will reach Darwin no later than 2020 (Freeland and Martin, 1985).

The cane toad was introduced into Australia to control pests occurring in sugar cane plantations. Whether it was the success it was hoped it would be is doubtful. The atti-

tude of the sugar cane industry was that anything that would assist the control of major economic pests was worth trying.

It is easy to be wise after the event but in reality many naturalists were extremely concerned about the proposal to introduce and release the toad in Australia, and it was a direct result of their protests that the belated (and temporary) ban on dispersal was imposed by the Federal government. Froggatt (1936) was one of the first individuals who suggested that the toad itself could become a pest.

ACCIDENTAL INTRODUCTIONS

To talk of 'introducing' an animal into a new area lacks some of the innuendo that we tend to associate with the expression of something being 'liberated': the inference of sexual freedom. But for cane toads perhaps it really is being liberated, because they tend to respond in new areas with sexual activities that ensure the perpetuation of their kind!

From time to time toads are accidentally transported (perhaps as stowaways) to a new area and released there. On other occasions they have been transported deliberately, perhaps for laboratory experimental use, but have escaped. In recent years there have been a couple of escapes that really created very grave concern.

In June 1974 a high school at Darwin imported from Queensland 20 cane toads for dissection by students in a biology class. It was reported that a teacher took the container of toads home to show his children, and at night left the container beneath the carport of his eastern suburbs home. Somehow it was knocked over and the toads escaped. Some of the toads were recovered in the immediate vicinity but 11 could not be found and so a huge and elaborate search was mounted.

Many Darwin residents argued that one of the unique attributes of their tropical city was that it lacked the cane toads whose dead bodies polluted the streets in tropical Queensland. Conservationists and farmers were outraged at the possible risk to native wildlife. The Darwin Conservation Society offered a reward of five dollars for each toad recovered. The President of the Northern Farmers' Association said that the reward should be

$1000 each and the money provided by the Department of Education because the toads had been imported by a school. The then Minister for the Northern Territory, Dr Rex Patterson, rejected the proposal, pointing out that more toads might be imported by people inspired to claim the high reward!

The most significant activities were undertaken by what was then called the 'Forestry, Fisheries, Wildlife, Environment and National Parks Branch' of the Depment of the Northern Territory. Today the relevant section is known as the 'Conservation Commission of the Northern Territory'. At the suburb where the toads escaped scarcely a leaf was left unturned, waterside vegetation where the toads might hide was burned and all available Rangers, biologists and technicians were brought in to take part. I was asked to provide help and in mid-July I visited the area, in time to learn that the Legislative Council had that day passed a Bill declaring

the cane toad a pest to be destroyed on sight, and imposing a penalty of $400 on anyone importing one without a permit. The debate had produced a great deal of criticism for the Department of the Northern Territory for permitting the toad into the Territory without any controls.

When I arrived I found schools and police stations displaying large reward signs (Figure 47), and there was no doubt that the matter was being taken very seriously. I had brought a tape-recording of the cane toad mating call. It was intended to enable searchers to recognise the call and so perhaps locate specimens calling at night. It was a serious suggestion but the media thought this the funniest idea they had heard for years. Cartoons and even poems appeared in newspapers, but the radio stations broadcast the call every night, and this helped recover one of the toads. What happened was a remarkable chain of events.

Some of the toads escaping from the

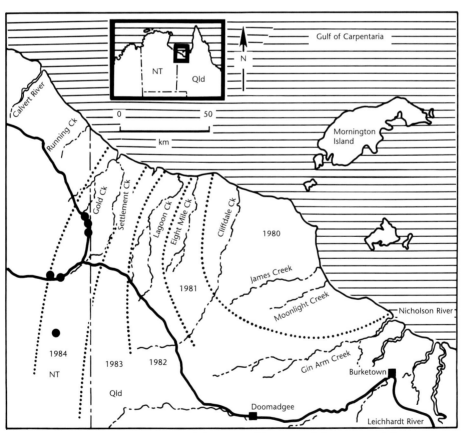

FIGURE 46 *Progressive expansion of the geographic range of the cane toad from Queensland into the Northern Territory. From Freeland and Martin (1985)*

– ESCAPED –

QUEENSLAND

CANE TOADS.

REWARD.

Six (6) Queensland Cane Toads are still at large in the Northern Suburbs. We believe they are in the Nakara, Adcock or Bradshaw areas. They are ground animals and are likely to be found in damp places such as the base of bushes, rockeries, edges of pools of water, under piles of rubbish, in laundries.

They could be dangerous to the Northern Territory environment. You can help to keep this animal out of the Darwin area by joining in a search of your neighbourhood.

If you capture one of these escapees (dead or alive), put it in a screw-top glass jar, and deliver it promptly to your School Principal.

FIGURE 47 *Poster produced in Darwin in 1974 following the escape of cane toads there.*

release site crawled through the grills leading to the huge storm water drains that prevent flooding in the downpours that characterise the northern wet season. They hopped along the giant pipes for at least 500 m to a water-filled culvert next to the local Community College. A couple of toads were recovered beside the culvert but one day workmen broke a high pressure water main and the car park was flooded. Some of the cars became bogged in the mud and it seems that a toad flushed out by the flood clambered onto a ledge or some part of a small Volkswagon. After the owner had been towed out of the mud she drove to her home several kilometres away. After she had parked her car the toad hopped out onto the lawn and, sitting beneath the sprinkler, recovered enough to start calling. The car owner's husband happened to walk out onto their upper-deck veranda, just as a radio station broadcast the recording of the call. He then heard the call from closer at hand, and went down and recaptured the toad.

It was extremely fortunate that the escape of the toads at Darwin occurred in the middle of the dry season. The high temperature and low humidity no doubt prevented breeding from occurring and all of the escapees were recovered or perished.

Within weeks of the accidental release of cane toads at Darwin, I was in my Adelaide office one day when a journalist from *The News* telephoned and asked 'What would you say if I told you that a consignment of cane toads has escaped at Perth Airport?'. I told him exactly what I would say, but he insisted that it was quite true. Later the same day I was asked by the Agricultural Protection Board of Western Australia to fly over to advise on the mopping up operation.

Cane toads have performed a vital role as experimental animals for use in biological and medical research in Australia. On this occasion one of three boxes freighted to Perth from Ingham in Queensland on 5 July was accidentally broken whilst awaiting collection at Perth Airport and at some time during the night of 8 July some of the toads escaped. The problem of capturing them was complicated by uncertainty about just how many toads had been in the crate. It should have

held 50 but at that time the people despatching the toads used to pop in a few extra to compensate for any deaths during the journey (it was more economical than handling claims), so it was any number between 50 and 56.

When the seriousness of the prospect of toads becoming established in the surrounding areas was appreciated the government came under great pressure. The Country Party members of the government represented beekeepers who were aware that cane toads eat bees and that in Queensland the honey industry had suffered badly from the presence of the pests. The beekeepers appealed to their Members of Parliament, and the members of the Beekeeping Section of the Farmers' Union joined busloads of volunteer searchers from secondary schools and the Teachers' Training College converging on Perth Airport. The Farmers' Union offered a fifty dollar reward for each recaptured toad, and the hunt was on in earnest.

Airports are designed for aircraft and people and car parking, and not with any thought for the recovery of escapee toads. Hence the search was complicated by physical factors such as the presence of a huge shrubbery in front of the passenger terminal (since cleared) and an escape route in the form of a complex system of underground drains for which the only plans were in Canberra.

The entire airport was searched several times. Freshwater ditches to the nearby marshes and Swan River were checked and the underground drains flushed through with the help of an airport fire tender. Fifteen toads were recaptured leaving perhaps five or six unaccounted for. Whether they perished in huge airport septic tanks or elsewhere is not known. For months the grounds were patrolled for any sign of toads or their spawn, but happily the cane toad did not become established.

The Darwin and Perth experiences should have acted as a catalyst to promote research into ways of controlling the toads or at least devising means of halting their western and southern spread from Queensland into the rest of Australia. But, unfortunately, when the furore died down nothing was done. Just five

Plate 40 *Tangled spawn of* Bufo marinus. *(G.C. Grigg)*

years ago (in December 1983) a toad reached Perth airport as a stowaway in a box of bananas.

There have been other alarms: on Christmas Eve 1976 someone broke into the Physiology Department at Sydney University and released 40 or 50 toads. The following year a vandal forced open the door of an exhibit at the zoo in Adelaide to release toads. In August 1980 a dead toad was found in Katherine Gorge in the Northern Territory. (A search failed to locate others and it was presumed to have stowed away upon a vehicle.) In May 1982 three cane toads were found as stowaways on an aircraft flying from Cairns to Darwin. Following that incident the Territory's Minister for Primary Production, Mr Steel, threatened jail and a $2000 fine for anyone found guilty of bringing in cane toads. He stated 'The devastating effects cane toads have on Australian wildlife is well known and the Territory is making every effort to ensure the creatures do not spread across the border'. In January 1984 three toads were stolen from the Salisbury College of Advanced Education near Adelaide. A major search was instituted involving the police force and National Parks and Wildlife Service officers. The State government offered a reward of $200 for each toad recaptured. One was found but the other two disappeared without trace.

In May 1983, a live cane toad was discovered on the island of Groote Eylandt in the Gulf of Carpentaria which lies close to the Northern Territory mainland of Arnhem Land. That one was assumed to have arrived amongst a ship's cargo, and a thorough search failed to locate any more.

FEEDING HABITS AND DIET

Most of the extensive literature on the diet of cane toads doesn't resolve whether the species is a selective feeder or not, simply because the articles report the identification of food items dissected from stomachs. Much of the early literature on the economic significance of cane toads falls into this category. There are dozens of brief reports stating the items recovered. For example, writing in the *Agricultural Journal* of Fiji, Bell (1940) reports: *A systematic examination of the stomach contents of toads was carried out during the period December 1938 to April 1939, to determine what constitutes the food of toads in their new habitat in North Queensland cane areas.*

Unfortunately it was not possible to collect toads in the few areas where Greyback beetles (Lepidoderma) *were flying in appreciable numbers, and accordingly this pest was not represented in the material taken from the toads' stomachs. However, a pleasing find was 15 'frenchi' beetles* (Lepidiota frenchi *Blkb.) in the stomach of one toad, and 13 in another, whilst reports come to hand that toads have gorged themselves on these beetles where the latter were plentiful. It was also gratifying to find that comparatively small toads (two inches to three inches in length) had eaten other sugar-cane pests, namely sugar-cane borer weevils* (Rhabdocnemis obscura *Bois.), and numbers of army worms* (Spodoptera exempta *Wlk.).*

More elaborate analyses on large samples of toads have set out to distinguish prey items according to whether they are economically beneficial, neutral or harmful. Pippet (1975) used this approach but was cautious about the classification, stating:

It will be appreciated that there are considerable difficulties in 'pigeon-holing' the majority of organisms in terms of economic importance. What may be termed a pest in one area, could well be of no significance in another.

Pippet's summaries are reproduced here in Table 11. They demonstrate that, in terms of number of prey items, the cane toads ate significantly high numbers that he considered harmful.

The consensus seems to be that toads are indiscriminate feeders, virtually eating anything that passes within their range of vision (see Plate 18). Because the toad has become extremely widespread in Australia it is worthwhile explaining just what does influence diet.

The major constraint on diet is the size of the prey. The cane toad lacks teeth and swallows its prey whole. (Other closely related species possess teeth but they don't chew either—the teeth have become a sort of optional extra.) Therefore the toad will attempt to eat something if it is small enough

to pass through its jaws. So it may eat a small snake or lizard whose body length is two or three times its own length; the significant matter as far as the toad is concerned is that the head and body of the prey is narrower than the gape of the toad's jaws. The expression 'to bite off more than you can chew' just doesn't apply because the toad will attempt to swallow the prey item whole!

The other factor that affects diet is where feeding takes place. If the toad lives close to water, it is likely to encounter more prey items that are dependent upon water for some stage of their life-cycle.

The activity of the toads is important. If they are calling and really giving their attention to locating a mate then they will not be feeding as well, and the stomachs will contain less food.

The stimulus necessary for capturing prey is also a very important factor, and here is where toads differ from native species quite markedly. All species of frogs that I've studied require the food item to move. If an insect or other object is motionless the frog won't accept it. But toads will eat dog's meat in bowls (and assemble each night at the back door beside the dog, waiting for his bowl of meat to arrive), and they have also been observed by Alexander (1965) in Florida to visit refuse tips to eat kitchen refuse such as corn on the cob, avocados, broccoli, rice, etc. They seem to use their powers of smell to assist them in locating food and also be prepared to accept non-traditional food items. But if objects pass within their field of vision whilst they are seeking food, they'll react in the normal way and swallow it. I've seen small stones and even lighted cigarette butts swallowed by toads, and they have been reported to try to eat bouncing table tennis balls, and to die from swallowing petals falling from strychnine trees. It scarcely seems possible that they taste what they are taking in but experiments by Forbes, Abbott and Hamre (1949) led to the tentative conclusion that toads have on their tongues the same four types of taste receptors as we do, so enabling the possessor to distinguish sweet from sour and salt from bitter.

We tend to be cautious about food items that are new to us. 'Taste this' someone may

TABLE ELEVEN: Economic importance of prey consumed by 350 *Bufo marinus* at Serovi Plantation, Popondetta, New Guinea, March to November 1973. In parentheses are similar data for Laloki Plantation for January to December 1973. (From Pippet, 1975)

TYPE OF PREY	NUMBER OF INDIVIDUALS		
	BENEFICIAL	NEUTRAL	HARMFUL
Annelida	3 (1)		
Mollusca			17 (1)
Arthropoda			
Crustacea		(2)	
Isopoda	2		
Myriapoda	2 (3)		(6)
Diplopoda			172
Scorpiones	(1)		
Arachnida	(12)	29	2 (1)
Insecta			
Blattoidea		16 (4)	
Isoptera			94 (26)
Dermaptera	20		
Orthoptera			87
Hemiptera	9 (6)	48 (47)	355 (82)
Coleoptera	190 (59)	228 (64)	402 (183)
Diptera	3	76	
Lepidoptera		105 (12)	348 (290)
Hymenoptera	3 (7)	954 (385)	
Vertebrata			
Squamata			
Scincidae	2 (1)		
	234	1456	598

say, offering a drink (perhaps an exotic fruit juice that you haven't tasted before). If you imagine now that you have just taken the glass and pretend that you are about to taste it, you'll find that you purse up your lips and bring the tip of your tongue forward. Go on, try it for yourself while you read on! What you are doing is using the tip of your tongue to taste rather than letting the fluid enter into your mouth only to find out too late that it is vile.

Toads don't share our caution, or at least don't give the slightest sign of doing so. If they really do use their taste buds it must be whilst they are capturing food with their sticky tongues. Having sniffed some of the sorts of insects the toads eat and found that they smell repulsive, I suspect that they would taste absolutely awful. Toads just cannot experience taste in the way that we do!

Toads will eat wasps and bees too and certainly get stung in the process. I've seen a cane toad stomach with the barbs of several bee stings completely penetrating its muscular wall.

The ability of toads to eat bees and, in fact, to visit hives especially to get a big feed there has been known to beekeepers for many years. Naturally beekeepers were amongst those who objected to the introduction of the cane toad in the first place and Lever (1944) confirmed their worst fears with his dissection of an immature female toad collected at Suva, Fiji (by a beekeeper) in January of that year. He noted that its stomach contained, 'A large mass of heads and wings of several hundred bees and two nymphal cockroaches. These were consumed near a beehive'. He went on with the observation, 'The writer is informed that a special barrier has to be erected in order to keep toads away from the alighting board of bee hives which clearly suffer badly from attacks by toads'.

But stories about cane toads always seem to become exaggerated and in *The Land* of 30 August 1946 we learn that delegates to an apiarists' conference in Brisbane were told that:

... when fully grown, these toads were capable of reaching up fully a foot, and would weigh anything from five to seven pounds with a tongue six inches long that swept bees off hive entrances in alarming fashion. In fact, one of the pests had been killed and, when opened up, was found to have 500 bees in its stomach.

The beekeepers visiting Brisbane may have been fooled, and from time to time the Brisbane apiarists have worried their colleagues from southern states with a few exaggerations designed to strike fear into their hearts.

But there is no doubt that the impact of the cane toad upon the honey industry was quite severe. The toads certainly did congregate around hives and devour bees in large numbers. Attempts to create barriers of wire mesh were time-consuming and, following recommendations published in the *Barbados Agricultural Journal* in 1939, the *Queensland Agricultural Journal* published in 1966 details of a collapsible table designed to raise hives 0.7 m above the ground and therefore out of the reach of the toads. Bearing in mind the vast amount of space needed in a vehicle to transport a couple of hundred hives from one place to another to coincide with the appearance of particular flowers, having to transport tables as well increased costs to beekeepers appreciably. It is experiences such as these that have led beekeepers to become a powerful lobby group urging governments to keep toads out of the southern and western states.

INTERACTIONS WITH OTHER ANIMALS

The word 'interaction' has always seemed to me a trifle technical, if not pompous, and I tend to avoid using it. But I think it's not *just* appropriate here but, in fact, perhaps the only way of saying literally 'having an action with each other'. I want to explore how other animals influence (or have an action on) cane toads and what action cane toads, have on other animals. At the most fundamental level: just how many different sorts of animals eat cane toads and, bearing in mind that the cane toads contain venom glands in their skin, just how many live to repeat the experience? And can cane toads live amicably with the native species? In conjures up images of Australia's various ethnic communities and their interrelationships (there we go again) with the native-born Australians. And that seems to me exactly how we should approach a study of the effects of the introduction of the cane toad upon the native fauna, upon domestic animals and upon humans.

Sifting fact from fallacy is sometimes a trifle difficult but in the first few years following the introduction of the toad into Australia some press reports were such gross exaggerations that their fallacy is obvious. Perhaps none was quite as ludicrous as that noted by S. Reimer (1959) in Florida following the release of toads there, describing them as:

monstrous toads which threaten housewives in their backyards, seize dogs by the head and hang on with a death-resulting grip, or attack and kill with their virulent poison the innocent neighbourhood cats.

When approached by a possible predator the toad is not agile enough to escape by leaping off, and so it commonly rolls its body towards its enemy, presenting the largest possible surface area, a ploy that presumably may deter smaller predators by appearing too big to be eaten.

But the skin of the cane toad is very poisonous with the greatest amount of venom located in a pair of large 'parotoid' glands located on the shoulders (Figure 48). Sections cut through these glands show large chambers of secretions (Figure 49). At the surface scanning electron microscope pictures reveal the apertures providing a passage for the venom in the glands to reach the outside (Figure 50).

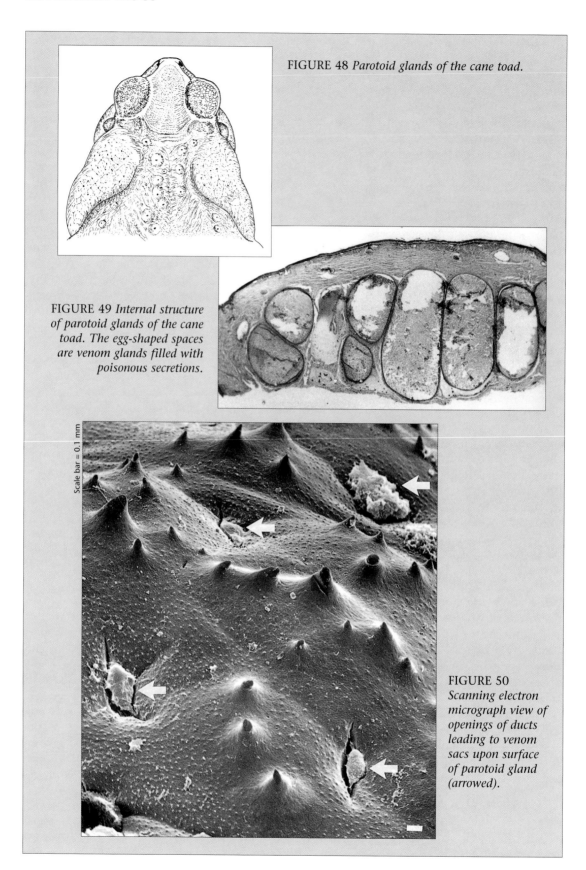

FIGURE 48 *Parotoid glands of the cane toad.*

FIGURE 49 *Internal structure of parotoid glands of the cane toad. The egg-shaped spaces are venom glands filled with poisonous secretions.*

Scale bar = 0.1 mm

FIGURE 50 *Scanning electron micrograph view of openings of ducts leading to venom sacs upon surface of parotoid gland (arrowed).*

If the parotoid glands are squeezed between thumb and finger some of the contents can be pushed out onto the surface and shown to be a milky, cream and rather viscous fluid. But under extreme provocation the venom doesn't just ooze to the outside but can be squirted violently in a spray for up to 1 m. Because the gland apertures are located all over the irregularly-shaped gland the spray travels in several directions at once. If the venom enters the eye of the predator (or a person holding it) the toad is likely to be dropped. As I can testify from personal experience, it produces excruciating pain followed by local irritation. It would certainly be enough to deter any animal that had picked a toad up intending to eat it.

If the toad is eaten, death of the predator seems inevitable in most species of mammals. Whether creatures, such as the mongoose or rat, that are reported to devour cane toads, eat every part of the body including the parotoid glands is not clear. Experiments with freshly killed specimens of the native frog *Limnodynastes dumelilii*, which has a poison gland on its calf, showed that everything was eaten except the calf glands (Crook and Tyler, 1981). If predators of the cane toad act similarly they are not exhibiting a tolerance to the venom—they are just being smart.

The secretions of the toad parotoid glands are not composed of a single substance but are best viewed as an elaborate cocktail. In each species there seem to be a few fundamental similarities but also unique biochemical compounds.

The cane toad includes in its parotoid secretions the chemical substances Marinobufagin, Telocinobufagin, Dehydrobufotenine, the catecholamines Adrenaline and Noradrenaline and also huge quantities of 5-Hydroxy-tryptamine (Tyler, 1987). Collectively they certainly provide the toad with defence against predators but it has been suggested that together these substances also provide protection against various kinds of microorganisms including bacteria and fungi that would otherwise grow on the toad's skin.

Just what effect the toads had upon Australian wildlife was never very clear and data accumulated painfully slowly. In fact the position was only clarified when Covacevich and Archer (1975) organised the distribution of 2500 circulars to their associates, veterinarians, schools, etc. The results are recorded in Table 12. In their discussion they considered the cane toad a potential hazard to local predators such as frog-eating snakes and native cats. They related the observation of Michael Archer who had kept a western native cat in captivity for three years.

During this time it was never observed to hesitate in attacking any small vertebrate. Although it often exhibited caution in the method of killing these prey, it never hesitated to bite. It was the result of a single bite (an accidental encounter) on an individual *B. marinus* that caused its death.

The effect of the toad upon domestic animals such as cats and dogs seems to depend upon whether the cat or dog eats the toad or if it simply 'mouths' it (that is, picks it up with its mouth and holds it there). If the toad releases venom from its parotoid glands then the animal holding it will suffer the consequences; if the toad doesn't release any venom and is dropped, then the cat or dog will have had a lucky escape. In the years following the release it may have been variability in the release of the venom that created disputes about whether or not toads constituted a hazard to pets. Otani, Palumbo and Read (1969) estimate that in Hawaii 50 dogs are killed each year from cane toad venom. Comparable figures for Australia are not available. Otani and his colleagues state:

small quantities of the toxin may cause profuse salivation, prostration, cardiac arrhythmia, pulmonary oedema, hypertension, convulsions, and death.

The rapid action of the venom and particularly the way in which respiration fails quickly, provides little chance of survival unless the animal can receive help promptly from a veterinarian.

Curiously, experiments with chickens showed that they were completely unaffected by toads: one ate 142 small toads in about an hour and another ate 83 (Buzacott, 1939). However small toads probably are not as toxic as larger ones.

In the period immediately following the

TABLE TWELVE: Records of native Australian animals killed as a result of eating or 'mouthing' cane toads.

	IDENTITY	REPORTED BY
REPTILES		
SNAKES	*Stegonotus cucullatus* —slatey grey snake	Covacevich & Archer (1975)
	Boiga irregularis —brown tree snake	"
	Pseudechis porphyriacus —red-bellied black snake	Covacevich (1974)
	Pseudechis papuanus	Pippet (1975)
	Acanthophis antarcticus —death adder	Covacevich & Archer (1975)
	Pseudonaja textilis —brown snake	"
	Notechis scutatus —tiger snake	"
LIZARDS	*Egernia major*[1]—land mullet	"
	Varanus spp.—goannas	"
	Varanus gouldii —Gould's goanna	Stammer (1981); W.J. Freeland (pers. comm.)
	Varanus panoptes	W.J. Freeland (pers. comm.)
	Varanus varius —lace monitor	Van Beurden (1980b)
BIRDS	*Dacelo gigas*—kookaburra	Covacevich & Archer (1975)
	Corvus sp.—crow	"
	Ixobrychus minutus —little bittern	Van Beurden (1980)
	Dupetor flavicollis —black bittern	"
MAMMALS	*Dasyurus geoffroii* —western native cat	"
	Dasyurus sp.	Pippet (1975)
	Sarcophilus harrisii —Tasmanian devil	Covacevich & Archer (1975)

[1] Reported as *Egernia bungana*

introduction of the cane toad into Pacific islands, deaths of humans resulting from eating that species instead of native species were reported (Rabor, 1952) and there have been accounts of illness resulting from people just handling them (Allen and Neill, 1956). These cases remain exceptional.

There has been considerable interest in finding out just which species of native animals are capable of destroying cane toads. The variety throughout the world is quite extensive (Table 13), but these records have been assembled over many years and it is clear that none can be considered a predator of any significance in the control of the cane toads.

Toads are often present in such huge numbers that it is hard to believe that they would fail to displace any native species that were around. But there rarely seem to be many native species competing with them. (This must be one of the reasons that the cane toad has been so successful in Australia.) In New South Wales, Van Beurden (1980a) concluded that there were one or two native species of frogs which were dependent upon ponds for breeding purposes. These he argued would survive coexistence with a few toads but would not survive competition against large numbers. Studies in New Guinea by Zug, Lindgren and Pippet (1975) resulted in the conclusion that native species there were not being displaced by the toad. But in the USA where the cane toad exists with other toads that are closely related to it, there is a better chance of competition taking place. In fact in a laboratory situation where the cane toad *Bufo marinus* shared accommodation and food with *B. americanus*, it eventually so dominated the feeding sessions that the *B. ameri-*

TABLE THIRTEEN: Animals reported in various countries to eat cane toads (eggs, tadpoles or adults) without adverse effects, or reported to kill toads.

	IDENTITY	REPORTED BY
INSECTS	*Orthophagus cuniculus* —dung beetle	Waterhouse (1974)
	Megadytes giganteus —water beetle	Anon
CRUSTACEANS	*Cherax* spp.—crayfish	Hutchings (1979)
	Euastacus hystricosus—crayfish	"
	E. valentulus— "	"
	E. suttoni— "	"
FISH	*anguilla* sp.—eel	Lever (1938)
	'top minnow'	"
	Kuhlia rubestris—jungle perch	Covacevich & Archer (1975)
	catfish	"
	perch	"
REPTILES		
CROCODILES	*Crocodilus porosus*	"
	Caiman latirostris —caiman	Pope (1955)
LIZARDS	*Varanus salvator*	Alcala (1957)
	Helicops sp.	Allen & Neill (1956)
SNAKES	*Leptodeira annulata*	Zug & Zug (1979)
	Amphiesma mairii	Covacevich & Archer (1975)
	Dendrelaphis punctulatus —green grass snake	"
TURTLES	*Chelodina* sp.	Gunn *et al.* (1972)
	—koel	Cassels (1970)
BIRDS	*Podargus strigoides* —tawny frogmouth	Covacevich & Archer (1975) Freeland (1985)
	—kite hawk	"
	—whistling kite	"
	—pheasant	"
	—ibis	Goodacre (1947)
	—crane	Covacevich & Archer (1975)
	—swamp hen	"
	—herons	Wingate (1965)
MAMMALS	*Hydromys chrysogaster* —water rat	Covacevich & Archer (1975)
	Rattus rattus—common rat	Adams (1967)
	—mongoose	Baldwin *et al.* (1952)

canus eventually withdrew from competition and four out of five of them died (Boice and Boice, 1971).

The significance of the competition experiments is that because the cane toad occurs naturally in Texas, and now has been introduced into other states, there is a real chance of competition taking place between these two essentially similar animals. Hence in Australia we know that it does remarkably well without any significant competitors, but it also seems to have that extra something for success when it does come up against them.

SCIENTIFIC OR ECONOMIC BENEFITS

Having related a tale of woe it is most important to redress the balance a little and point out the ways in which the cane toad has been of benefit in Australia. The first of these involved use in a new clinical test.

In 1947 Galli-Mainini of Argentina reported that if a male of the South American toad *Bufo arenarum* was injected with a

gonad-stimulating substance, it would release spermatozoa which, because the testes were closely associated with the kidney, very quickly (within three hours) ended up in the urinary bladder. The gonad-stimulating substance occurred in the urine of pregnant women and in the following year Galli-Mainini reported that many tests had demonstrated that the technique provided a new, accurate, rapid and economic test to confirm pregnancy, even at early stages.

PLATE 42 Bufo marinus. *(P. Kempster)*

PLATE 41 Bufo marinus *habitat on Rainbow Beach, Queensland. Tadpoles were in the water in the foreground. (M.J. Tyler)*

PLATE 43 *A male* Bufo marinus *attempting to mate with a female who has been dead for several days.* (*M.J. Tyler*)

Galli-Mainini's observation led other workers throughout the world to examine the response of many other species of toads and frogs, and they reported that the test seemed to work in all of them. The only difficulties they experienced were in practical matters such as distinguishing male toads from females. By 1948 Lima and Pereira had shown that the cane toad was a suitable subject.

By 1949 the cane toad was being used routinely for pregnancy tests in Australia (Bettinger and O'Loughlin, 1950). The technique was simply to inject with a syringe 5.0 ml or 10.0 ml of urine from the subject into the lymph sacs that separate the loosely fit-

ting skin of the toad's flanks from the underlying muscles. To ensure the greatest possible concentration of the hormone in the urine, the patient was asked to collect the sample in

PLATE 44 Bufo marinus *killed on the road.* (*M.J. Tyler*)

125

the morning after a night without taking any fluids.

After injection the toad was left for three hours and then a small pipette of 0.5 ml or 1.0 ml capacity was inserted into the cloaca of the toad and moved about gently. Soon a drop of toad urine could be collected (often a flood). The urine was then examined under a compound microscope for a sign of any motile spermatozoa. Presence of spermatozoa indicated that the patient was pregnant. If after three hours no spermatozoa were detected the toad urine was re-examined after six or seven hours. If the result was still negative the test would be repeated in a second toad, and a negative test then taken to indicate that the patient was not pregnant.

The cane toad was used in pregnancy tests for about 15 years, but the test has now slipped into medical history, made obsolete by a biochemical test on urine that doesn't require toads and gives a positive result in three minutes.

The 1950s and 1960s also saw the cane toad being used in physiology and pharmacology. Routine assays of acetylcholine, the chemical transmitter at the nerve–muscle junction, were performed using a long slip of muscle dissected from the wall of the abdomen of the cane toad. It was also found that the heart could be maintained outside the body and the action of drugs observed at close quarters. The cane toad continues to this day to be significant in laboratory studies in terms of the contribution that it makes, but certainly not to the same extent as 25 years ago.

Probably the largest use of all has been in school, college and university biology classes, where toads have provided a simple vertebrate animal with which to study fundamental anatomy. The suppliers have been private dealers and Apex clubs in Queensland. In South Australia 557 were imported under licence in 1982 compared with 11 000, twenty years earlier.

Amongst the more bizarre uses was an attempt (apparently unsuccessful) to prepare a chicken food from the skinned bodies of toads in the Philippines (Rabor, 1952), and more recently a venture to prepare cane toad skins for tanning. Output of 150 000 skins a year from a factory at Nambour, Queensland was anticipated. The Queensland Government's wedding gift to Prince Charles and Princess Diana was a book bound in cane toad skin.

Recently I was shown an article in an Australian magazine claiming a new use for cane toad venom: used as an anaesthetic in dental surgery it is more effective than cocaine and stops bleeding of gums after tooth extractions. But I think I'll stick to the old techniques.

LEGISLATION AND CONTROL

Fauna legislation sets out to do one of two things: either to protect a species in some way, or else to prohibit people from taking particular species to certain areas. Shortly after the arrival of the cane toad into Australia the Commonwealth government imposed a ban on its release into new parts of Queensland. Following the lifting of that ban there was a period when people were found to be killing toads, and so the possibility of protecting the cane toad by Queensland legislation was explored.

But 15 years after the introduction of the cane toad into Australia the position had changed quite radically. I have not been able to trace the history of legislation within the States but in 1953 workers in New South Wales began examining native frogs and lizards for pregnancy testing purposes, because the cane toad 'is prohibited to other states as a noxious animal'. Western Australia declared the cane toad 'vermin' during the period but I have not traced the existence of legislation in other States until much later. Certainly in 1963 a South Australian government advisory committee rejected a suggestion to licence imports and no controls existed in South Australia until 1974.

Currently very strict controls apply to the importation of cane toads into Western Australia, South Australia, Northern Territory, Victoria and Tasmania. Permits are required to import toads for use in experimental and teaching laboratories, and severe penalties are prescribed for breaches of the regulations.

Less than three years after the introduction of the toad into Australia, the writing was on the wall that steps would have to be taken to constrain the toad population outside the areas for which they had been intended.

Is it too late to do anything? With a range in Queensland of over half a million square kilometres and expansion at a rapid rate, the prospect is a forbidding one. But the cane toad is an exotic species and there are a few useful avenues for research. Obviously we need to develop a control system that is absolutely specific for the cane toad and will leave the native species unharmed. That pipe-dream will depend upon having a more detailed knowledge of the toads' biology and of the native species as well. And here we have a stumbling block because we don't even know how many native species live on our continent—every frog expert in Australia will agree that not enough detailed exploration has taken place in the more remote northern areas, particularly those which the cane toad will invade shortly.

There remain other possible control measures. Heatwole and Shine (1976) suggested the possibility of transmitting diseases to cane toads by means of mosquitos. Van Beurden (1980) went further with observations of the mosquito *Mimomyia elegans* feeding on them, indicating the possibility that if a specific disease of cane toads exists (perhaps in South America), here was a mosquito that could transmit it.

For sheer ingenuity Waterhouse's (1974) proposal takes first prize. Concerned at the fact that the cane toads have been eating dung beetles imported to help bury the dung of cattle, and in fact waiting at the dung pats for the beetles to arrive, he suggests a 'possible countermeasure is the introduction of certain giant, heavily armoured and immensely powerful dung beetles of the genus *Heliocopris*. Some are almost the size of a golf ball and cannot be retained in the closed hand, so powerful are the digging motions of their legs. They fly principally at dusk or dawn, when toads are active, and it seems likely that if a toad swallows one of these beetles whole (as is the toad's habit), the beetle would be strong enough to break out through the toad's body wall'.

But at present the only real control of the cane toad in Australia, and probably elsewhere in the world, is performed unwittingly (see Plate 44).

In November 1984, almost 50 years after the liberation of the cane toad in Australia, a workshop was held in Brisbane at the headquarters of the Queensland National Parks and Wildlife Service. It was convened to explore options for the control of the toad and involved contributions from a number of Australian and state authorities. As a result of the meeting CONCOM, the organisation of state and commonwealth conservation and wildlife departments, funded a two-pronged program. One segment involved an examination of the toads at the point of their penetration into the Northern Territory, whilst the other an examination of diseases affecting the toad.

For 10 years there has been considerable research upon the cane toad funded variously by some of the State governments, but principally by the Australian government. There have been several facets to the study, including ecological investigations to clarify the events that influence dispersal, diet and population size.

The search for a disease that might control toad numbers was extended to the animals' native home of Brazil and Venezuela. Currently the focus of the research in Australia is for a viral or other disease and clarification of the impact of the toad upon the Australian fauna. The CSIRO Division of Wildlife and Ecology in Canberra is coordinating the research with a budget of $1.25 million spread over three years.

CHAPTER ELEVEN

THE SANDHILL FROG

One of the greatest joys that can be experienced by a naturalist is the actual moment of discovering a new species. It is difficult to express, but to pick up an organism, look at it briefly and realise in a second or two that you are looking at something new, can really cause the adrenalin to flow!

On most occasions the reality of discovery is a trifle different. In a continent such as ours the chances are that most new species of vertebrates discovered differ only slightly from known ones, and so it may require a good deal of subsequent study before the certainty of the uniqueness of the find can be confirmed or even realised for the first time. But from time to time a creature is found that is so different from other known forms of life, that the realisation is virtually instantaneous. The sandhill frog was, for me, such an experience.

In 1974, I made a short visit to the Western Australian Museum in Perth to examine the collection of preserved frogs there. At that time it was housed in far from ideal conditions in the basement of a former prison beside the main museum building. The lighting was dim and I had difficulty in seeing the contents of the jars that filled the stacks of shelves extending from floor to ceiling.

PLATE 45 *Sandhill frog* Arenophryne rotunda *from Shark Bay, WA. (M. Davies)*

FIGURE 51 *The sandhill environment at Shark Bay, Western Australia.*

The collection was organised in an alphabetical sequence, and on the second day I arrived at the shelf housing the *Pseudophryne* species, better known as toadlets. As I picked up each jar in turn I came upon one labelled *'Pseudophryne guentheri* ?new sub-species'. But although I had to agree that the colour pattern of the frogs was vaguely like *P. guentheri*, the shape seemed different. I took the jar away and examined the specimens closely with the aid of a microscope. I could not contain my excitement: this was not just a new sub-species; it was not even a *Pseudophryne* species but a frog totally unlike any known in Australia.

Examination of data in the register revealed that the series of 10 specimens had been collected by Alex Baynes and Tom Smith at Shark Bay in August 1970. Alex was at the Museum at that time so I went to him to learn more about these strange frogs. He told me that they had fallen into traps that he had excavated to catch small mammals. But the most intriguing point that he made was that the site was in coastal sandhills where no fresh water could lie for long upon the surface.

Back in Adelaide I studied the frogs more closely. The largest had a total length of just 31.4 mm and contained white eggs 3 mm in diameter. The absence of any black pigment upon the surface of the eggs suggested that they would be laid away from direct exposure to light, whilst their size in relation to the size of the female, indicated that the young could perhaps live upon yolk reserves (direct development), and therefore not have a free-living tadpole stage.

The hands of the frog were unusual in that the outer finger was very short and it and the remainder were broad (see Figure 35 on page 90). This feature suggested that the hands might be used for burrowing, whereas almost all other burrowing frogs in Australia burrow using only their feet.

The internal anatomy of the species was of a form that hinted at a close relationship with the turtle frog *Myobatrachus gouldii* of Western Australia, and also with *Pseudophryne* (with which it had first been confused). I named it *Arenophryne* (Sandhill toadlet) *rotunda* (rotund) (Tyler, 1976c).

But preserved frogs have little appeal.

129

FIGURE 52 *False Entrance Well Tank at Shark Bay, Western Australia. Type locality of* Arenophryne rotunda.

What I wanted to do was to go to Shark Bay and watch the frogs in their natural environment. The first opportunity arose in 1979. So I followed the requirements and applied for a permit to collect frogs in Western Australia. In due course I received the permit, which informed me that I could collect any species except *Arenophryne rotunda* because it was considered endangered! It seemed that on 3 February 1978 *Arenophryne rotunda* had become the first frog in Australia to be protected by State government legislation. Why remains a mystery. I therefore sought (and fortunately obtained) a special permit to study this protected species.

In August 1979 Dale Roberts from the University of Western Australia, Margaret Davies from the University of Adelaide and I travelled to Shark Bay to search for the frog.

Shark Bay lies within an area of strangely shaped peninsulas and islands jutting out into the Indian Ocean. The names of the sites are equally unusual—the classic being 'Useless Loop' which sounds like a gynaecological disaster.

We arrived late one afternoon at Carrarang Station and were depressed at what we saw: just a vista of sand dunes and low heathland (Figure 51). In our experience frogs did not exist in such country. There had to be a mistake.

Our final destination was named 'False Entrance Well Tank'. We drove to the site called False Entrance and there, in the distance, could see the derelict concrete tank, once supplied from a nearby bore (Figure 52). We stopped and looked around, each of us privately thinking that there was no way in which any species of frog could survive there. But on the sand in front of us lay a couple of sheets of corrugated iron, for frogs a great, man-made refuge. I lifted one of the sheets and there, hiding away in one corner, were three young *Arenophryne rotunda*.

Over the following two days we learned a great deal about the sandhill frog and how it survives. It turned out that the frogs normally burrow into the sand at dawn and emerge at night. The significance of the burrowing is simply that although the sand at the surface is bone dry, at a depth of 10 cm or so it is moist. The moisture is fresh water (1.5 per cent by weight and with less than 10 parts per million (ppm) of dissolved salts. This, for example, compares with a range of 225–698 ppm and a mean of 390 ppm in the Adelaide tap water). Because the sand grains are large the tension they apply to the surrounding water is reduced, and it is possible for a buried frog to absorb water through its skin whenever it is in contact with the moist sand.

Early one morning we took photographs of frogs that we had captured the previous night. We placed them upon the sand and one of them tried to run away. We noticed that as it waddled, the scraping of its little hands and feet left tracks upon the surface. As the sun rose above the dunes, so these tracks were thrown into relief and we spotted many others upon the dunes (Figure 53). Where the tracks ended there was a little crater marking the spot where the frog had buried itself. All that we had to do to collect specimens was to follow their tracks and dig below the craters at the end of them!

The method of burrowing employed was quite unlike the normal pattern of backwards shuffling used by desert frogs. Instead the sandhill frogs burrowed headfirst. At the chosen spot they simply dropped their heads and dived beneath the surface. Initially they descended at 45° looking rather vulnerable with their little bottoms and hindlegs projecting into the air. But as they got deeper they were able to gain a purchase with the feet and push themselves in further.

In the dunes behind False Entrance Well Tank the sand surface is criss-crossed with the tracks made by the frogs. One afternoon, to get an idea of the number of frogs around, we brushed away all tracks in a couple of areas 5 m broad. When we examined the areas the following morning we found that 21 frogs had crossed one span, and 19 had crossed the other.

Working out how far the frogs travelled at night was a simple process. With a tape measure we just measured the distance between the blip on the sand left by an emerging frog and followed the tracks in the sand to the spot where it had buried itself. We found that it was easiest to follow the tracks at dawn and

dusk when the low angle of the sun highlighted any marks on the sand.

The distances travelled by the frogs ranged from 8.8 m to 27.6 m with an average of 14.8 m. The upper limit is quite remarkable for frogs in which the smallest adult known has a head plus body length of only 21 mm whilst the largest measures 36 mm.

To find what the frogs were feeding on, we went out with our torches one night, treading softly and hoping to find some in the act of eating. After half an hour we were rewarded with the sight of a little frog peering down into the conical cavity marking the entrance of an ant nest.

Ants are certainly the most abundant insect life upon the dunes. We confirmed that ants are a significant component of the frogs' diet by keeping two frogs in a plastic bag, and collecting their faecal droppings.

The one thing that we failed to achieve

FIGURE 53 *Tracks upon the sand at Shark Bay, Western Australia made by* Arenophryne rotunda.

during our brief visit was to find out anything about reproduction. In our paper we suggested (Tyler, Roberts and Davies, 1980) that the sandhill frog did not breed in water, and supported the suggestion that it had direct development.

Being based in Western Australia, Dale Roberts had easiest access to Shark Bay, and could study the frogs at intervals throughout the year. So Dale made a special study of the species and was eventually rewarded for his efforts (Roberts, 1984). It transpired that the males and females form pairs at some time in the winter or early autumn (July to November). They burrow together and remain below the surface for at least five months. As the sand dries out the frogs descend deeper and deeper, following the boundary of the moist layer to a maximum depth of 80 cm.

Mating takes place underground, so what actually occurs there remains a secret. The female lays 6–11 separate, creamy white eggs in circular capsules. The diameter of each egg averages 5.5 mm. The development of the embryo takes place entirely within the capsule. In Dale's laboratory baby frogs emerged from the capsule 10 weeks after the collection of eggs.

In September 1979, J. Rolfe and his colleagues from the Western Australian Wildlife Research Centre set pitfall traps near Kalbarri and Cooloomia Homestead near the mouth of the Murchison River, 250 km south of Shark Bay. Twenty-two sandhill frogs were recovered from these traps, representing an extremely significant extension to the known range of the species. Whether the species exists continuously along the intervening coastal sandhills, or in smaller separate populations is unknown, but the need for the species to be protected by special legislation must now be questioned.

Roberts (1985) attempted to estimate population numbers at Shark Bay and he compared the figures he obtained with those of other species in Western Australia. The technique employed involved using what are termed 'drift fences' and 'pitfall traps', that is,

PLATES 46–48 *Sandhill frog burrowing head-first.* (M. Davies)

PLATE 49 *M.J. Tyler standing at the end of a track made by the sandhill frog. (M. Davies)*

simply erecting a fence and placing pits at intervals in front of it. It is commonly used in reptile surveys and, less frequently, for frogs as well.

Dale erected 300 m of fence made of fly wire and sunk plastic pots as pit traps at intervals 4.5 m apart. He left the pits open for four or five nights by which time the capture rate had fallen close to zero, implying that almost all frogs in the area had been caught.

The population density of sandhill frogs was high: 277 per hectare, and with a collective weight (biomass) of 530 grams/hectare. When this figure is compared with the probable geographic area occupied by the species (5300 km²), it is clear that there are a lot of sandhill frogs around.

Although *Arenophryne rotunda* is an unusual and, in many ways, unique frog, there are other Australian frogs which lay eggs away from water and one (the turtle frog *Myobatrachus gouldii*) which burrows with its hands. In the original description I had suggested that the nearest relatives of *Arenophryne* were *Myobatrachus* and *Pseudophryne*.

Myobatrachus occurs south of the area occupied by *Arenophryne*, whilst *Pseudophryne* occurs both to the north and south. There have now been two more studies of *Arenophryne*: first, a more detailed examination of anatomy by Davies (1984), followed by a biochemical study (Maxson and Roberts, 1985). Both have confirmed the original suggestion, that *Myobatrachus* is the favourite candidate for closest relative.

In 1983 the ABC prepared a TV documentary *The man who loves frogs*. In suggesting to the film's director, Dione Gilmour, which species offered the greatest potential, I sang the praises of the sandhill frog and the aridity of the area in which it lived. What was needed was film of the frogs running upon the dry surface, making tracks, as well as their burrowing behaviour. So in August of that year three four-wheel drive vehicles set off from Perth on the 725 km trip to Shark Bay. Our group were Dione, a cameraman, Don Hanran-Smith, a sound recordist, Graeme Watson and Angus Martin from the University of Melbourne, with Margaret

Davies and myself from Adelaide. We reached Shark Bay the following day, set up camp at the now familiar site of the Well Tank at False Entrance, and still had time to head into the dunes to see the tracks of the frogs.

When we awoke the following day there were threatening clouds on the horizon and, at 9.00 a.m., a heavy squall battered us and wiped out all traces of the tracks upon the dunes. The team filmed the local scenery but the weather deteriorated further and by nightfall the rain was teeming down and the winds had become gale force. The tent housing Davies, Watson and Tyler was as efficient at excluding rain as a kitchen collander, and we were reduced to bailing water through the entrance, and simultaneously holding the tent down at the corners to stop it collapsing. Almost the entire night was passed in this fashion.

For two days we were hit by a series of squalls and storms; boots filled with water, all our clothes were damp, bedding was soaked. Somehow the film crew were able to take a few essential sequences, but for most of the time the sand remained hard, wet and compacted and it was with difficulty that they managed at last to film tracks—the main reason for the visit. Eventually we pulled out, to find the road back 15 cm deep in water. So perhaps a species with tadpoles could survive there after all!

CHAPTER TWELVE

GASTRIC BROODING FROGS

THE STATE OF KNOWLEDGE

Few frogs in the world have become as well known as the gastric brooding frog *Rheobatrachus silus* of the Conondale Range in south-east Queensland. Following its discovery in 1973 it was found that females brood their young within the stomach, and give birth through the mouth (Corben *et al.*, 1974). Photographic evidence of the birth process (Tyler and Carter, 1981) was reproduced in newspapers and magazines throughout the world, and attracted considerable attention. Then in 1984 a second species of gastric brooder (*R. vitellinus*) was discovered in central coastal Queensland (Mahony *et al.*, 1984; McDonald and Tyler, 1984). Tragically both species have disappeared, and it is conceivable that *R. silus*, at least, is now extinct. The location of their habitats is shown in Figure 54.

Despite great interest very little work was possible on *R. vitellinus*. The two papers cited above provided detail of the type specimen, including osteology and chromosome number, and details of oral birth. Leong *et al.* (1986) described the structure of the brooding stomach, whilst Hutchinson and Maxson (1987b) reported the relationships of *Rheobatrachus* species to other leptodactylid frogs. This appears to be the sum total of published information .

HABITAT AND BODY FORM

Both species inhabit boulder-strewn, fast flowing creeks in rainforest (Figure 55). The water is cool and usually clear, and the frogs hide away beneath or between the boulders in the current or in backwaters.

Physically the frogs are not remarkable. Both are brown or slate above; *R. silus* is pale cream below whilst *R. vitellinus* has extensive brilliant yellow patches and dark brown markings. In both species the skin is smooth and extremely slimy (Figure 56). The limbs are muscular and there is extensive webbing between the toes. *R. vitellinus* is larger than *R. silus*, females reaching a maximum length of 78 mm compared with 54 mm in *R. silus*.

DEVELOPMENT AND BIRTH

The early stages of the life-cycle are unknown. Eggs in gravid females were found to be up to 5.1 mm diameter. Such large-yolked eggs are characteristic of species lacking the need to feed during their period of development. But although females and males have been observed in amplexus (in November through to January), the laying and fertilisation of the eggs and the swallowing of the eggs by the female has never been witnessed. Whether they are laid in the water or on the banks is unknown.

The number of ripe eggs in gravid females (approximately 40) far exceeds the number of juveniles ever found to occur in the stomach (21–26). Thus either the female fails to take up all of the eggs laid, or else perhaps the first few swallowed are digested. It's worth noting that Ehmann and Swan (1986) discovered that not all of the tadpoles of *Assa darlingtoni* found their way into the males' hip pockets: there was wastage.

Initially the tadpoles of *R. silus* lack pigment cells in the skin because living in the stomach (out of the sight of predators) they have no need for camouflage. The only colour visible is that of the yolk sac which is

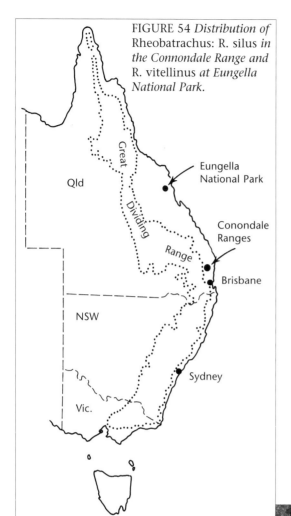

FIGURE 54 *Distribution of* Rheobatrachus: R. silus *in the Connondale Range and* R. vitellinus *at Eungella National Park.*

a pale creamy yellow. Skin pigmentation is acquired progressively as they develop. Information on tadpole development has been gleaned from a batch regurgitated by the mother, and successfully reared in shallow water.

Tadpoles at various stages of development are shown in Figure 57, whilst the differences between *R. silus* tadpoles and those of any frog species with a free-living tadpole are listed in Table 14. The fundamental difference influencing body form involves food sources,

FIGURE 55 (Bottom): *Typical habitat of* Rheobatrachus silus *in the Connondale Range, Queensland.*
FIGURE 56 (Below): *Adult female* Rheobatrachus vitellinus.

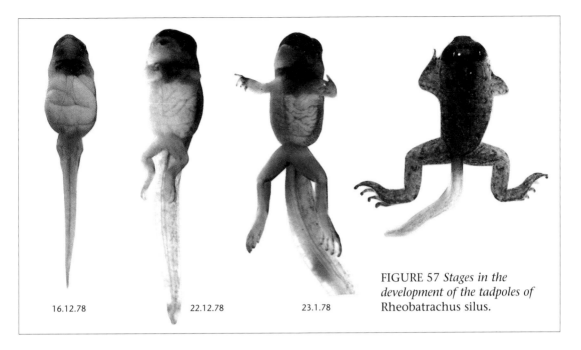

16.12.78   22.12.78   23.1.78   FIGURE 57 *Stages in the development of the tadpoles of* Rheobatrachus silus.

for whereas normal free-living tadpoles browse almost perpetually and so need to be active, *R. silus* is largely passive and the opportunity for movement within the confines of the mother's stomach decreases as the tadpoles grow.

Tadpole development In *R. silus* takes at least six weeks. The continuously warm laboratory conditions did not mimic the cooler and fluctuating temperatures that occur in the field, and this figure may be regarded as the minimum and probably conservative.

As the tadpoles grow and increase in bulk and weight, so the mother's stomach becomes more and more distended, progressively filling the body cavity. In the latter stages it has been demonstrated by X-ray that her lungs collapse, and so she would need to rely

more upon gas exchange through the skin. But, surprisingly, the mother remains active, despite the fact that the presence of babies in the stomach shifts her centre of gravity and changes her buoyancy, so that she is able only to hang vertically in the water when at rest.

In the final stages of their development the tadpoles have fully developed limbs and their tails become like narrow streamers. In the container in the laboratory these final stage young lay upon their backs as commonly as upon their fronts (see Figure 58), a habit no doubt prescribed by the cramped conditions that they would experience within the stomach.

The degree of distension of the body of the female certainly is remarkable. Figure 59 shows the young that had emerged from the

TABLE FOURTEEN: Comparison of the tadpole of *Rheobatrachus silus* with normal (free-living) tadpoles of other species.

FEATURE	NORMAL (FREE-LIVING) SPECIES	*RHEOBATRACHUS SILUS*
Labial teeth and horny beak	present	absent
Intestine	formed at early stage of development	formed at late stage of development
Tail	strong	weak
Arms	develop beneath body wall	develop outside body wall
Movement	almost constant	largely passive

FIGURE 58 *Late stage tadpole of* Rheobatrachus silus *lying upon its back.*

accompanying female and, when that photograph was taken it was not appreciated that one further baby remained within her stomach. The distension required to accommodate all of that group can be imagined. The distended stomach after birth of the young is shown in Figure 60.

At the completion of their metamorphosis there is very little variation in colour, length or size of the members of a clutch. Tyler (1983) reported that a clutch of 21 varied from 11.9 to 12.9 mm, and 0.248 to 0.305 g.

Initial observations suggested that the birth process was an extremely rapid event, for the females disgorged their entire comple-

ment of young by the simple act of propulsive vomiting: a startling sight. But when not handled or in any way disturbed the births are widely spaced out over a period of as long as one week. The mother rises to the surface of the water, dilates her oesophagus, permitting a baby to pass up into the buccal cavity where it sits upon the tongue prior to stepping out through the widely opened jaws (Figure 61).

THE INTERNAL ENVIRONMENT

It has been documented that at the time that the female is ready to ingest the fertilised eggs, her stomach is perfectly normal, and no different from that of any other species of frog. The frog stomach operates, in just the same way as the human stomach: in response to the swallowing of food, the stomach lining secretes quantities of hydrochloric acid and, in this acid medium, the digestive enzyme pepsin is produced to bring about the breakdown of food.

The first question facing the investigators of this strange phenomenon was to find out just why the young were not treated as food items and so digested. As indicate above, it is quite possible that the first few to be swallowed are indeed digested. Nevertheless acid production is switched off before it can harm the remaining members of the clutch.

Detailed accounts of the research that led

FIGURE 59 *Female* Rheobatrachus silus *with young. At the time that the photograph was taken it was not appreciated that another baby remained within her stomach. Reprinted with permission from 'converting a stomach to a uterus: the microscopic structure of the gastric brooding frog* Rheobatrachus silus', *by J. Fanning, M. J. Tyler and D.J. Shearman,* Gastroenterology, *Vol. 82, p. 63. Copyright 1982, the American Gastro-enterological Association.*

to the discovery of acid inhibition have been provided by Tyler (1983, 1985d). It appears that the jelly surrounding each egg contains a substance called prostaglandin E_2 (abbreviated to PGE_2): a substance with the known capacity to switch off the production of hydrochloric acid in the stomach. It is presumed that this source of PGE_2 dampens acid production throughout the early embryonic life of the developing eggs.

When the tadpoles of *R. silus* hatch from the eggs it seems that they too secrete PGE_2 to maintain the stomach in a non-functional state. Most probably the source of the PGE_2 is within mucus secreted in the tadpole gill arches. In the tadpoles of species which feed, small particles of food are trapped in cords of mucus in the pharynx and are then carried in a continuous chain down the oesophagus into the gut. In tadpoles of species such as *Rheobatrachus* which live upon yolk reserves it is likely that the pharynx still produces cords of mucus but that, because there is no open passage to the gut, they are excreted to the outside. My first awareness of this phenomenon occurred when I spotted fine filaments in the aquarium water surrounding some prematurely released tadpoles. To me they looked like strands of fluff. But when I examined a tadpole in a little dish under a microscope, I could see that mucus was being liberated from the exit from the gills (the spiracle), and the material in the water was simply

FIGURE 61 *Oral birth of* Rheobatrachus silus.

mucus cords.

Studies by Wassersug and Karmazyn (1984) on the common American bullfrog *Rana catesbeiana* have shown that its tadpoles also liberate PGE_2 and that production halts at the moment that the tadpole reaches the final stages of development into a frog. Noting that the development of the stomach does not occur until this precise moment, those authors suggest that one of the effects of PGE_2 is to inhibit tadpole stomach development.

In *Rheobatrachus silus* portions of the stomach wall of the female carrying babies undergo a massive and radical change, dilating and becoming as thin and as transparent as a plastic bag. It is curious that, in contrast, the stomach of *R. vitellinus* is not changed at all and remains normal throughout the period of gestation (Leong, Tyler and Shearman, 1985).

Apparently there are few other effects to the mother. But one of significance is that the part of the gut below the stomach is paralysed throughout the gestation period. This conclusion was reached from the observation that the intestine of *R. silus* brooding females contained undigested food material. If normal peristaltic movements of the gut took place after the entry of the babies into the stomach, the gut would be empty. The implication was that PGE_2 was in some way responsible for producing the paralysis.

It was not possible to test the role of PGE_2 in the production of gut paralysis on *R. silus* simply because of their rarity. We therefore selected the poor old cane toad *Bufo marinus*

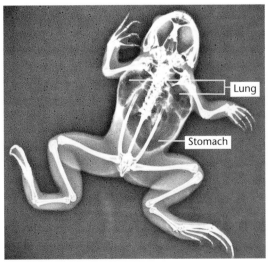

FIGURE 60 *Distended stomach of a female just after birth of the last of her young.*

as a subject. The first step was to examine the effects of PGE_2 on small portions of toad intestine maintained in a living state outside the body in special chambers. These experiments revealed that PGE_2 produced relaxation of the muscle wall of the stomach (de la Lande *et al.*, 1984). However it is not always possible to assume that results obtained from the responses of isolated portions of organs indicate what happens in an intact living animal. Further experiments were therefore devised to work out whether PGE_2 delayed the emptying of the stomach and of the passage of food through the gut. They involved feeding toads pellets of ground chicken liver containing radioactive technetium. PGE_2 was mixed with the pellets fed to one group of toads and omitted from others acting as controls. After periods ranging from three to six hours the toads were killed, the entire gut was removed and separated into portions, and the amount of technetium within each portion was checked.

The results showed that the release of food from the toad stomach and its passage down the gut was delayed in the presence of PGE_2 (Taylor, Tyler and Shearman, 1985). This result indicated that PGE_2 had a variety of functions and also that most of the physiological results we noted could be attributed to the secretion of this one chemical substance by the tadpoles.

To summarise the changes to the stomach of *R. silus* during the period of brooding, hydrochloric acid and pepsin secretion are halted and both the capacity of the stomach to discharge its contents, and of the intestine to move material along its length are affected. Because of its ability to stretch substantially the stomach is converted to a womb.

Nothing is yet known about how, or in what form, the young excrete waste material, or how they are able to obtain the oxygen they require to sustain life. But there are always questions unanswered. Solve one and someone will come up with another. Thus the difficulty is often just to decide which components are essential, and which ones fall into the fascinating but non-essential area. For example, whenever I give a seminar on the gastric brooding frogs there will be questions raised like 'Have you tried to find out whether males can carry babies in the stomach without digesting them?' The answer is 'no'. Even at the best of times adults and babies never have been sufficiently plentiful to make it possible. I do know that, except for being slightly smaller, the male stomach is in no way different from that of the female. Males don't swallow the eggs but, if they were force-fed, there seems no reason why their stomach should not be converted to a womb. But, to me, this sort of information lies in the domain of 'fancy-that' research and unlikely to gain a high priority even if adult males were plentiful.

It is now impossible to investigate anything in *R. silus* simply because the frogs cannot be found. And *R. vitellinus* has not been seen for several years. It is for this reason that physiological investigations shifted to other species. But this step can also be justified on the basis that frogs don't seem to differ from one another that much, so that at least in some circumstances it should be possible to extrapolate the results from one species to another.

CHAPTER THIRTEEN

FROGS AS ENVIRONMENTAL MONITORING ORGANISMS

ABNORMALITIES AND THE ENVIRONMENT

In recent years it has been noticed that the eggs and tadpoles of frogs are sensitive to a wide variety of environmental pollutants, which commonly result in the production of abnormalities of soft and/or skeletal tissues. In many of the reports the association of a particular pollutant with one or more abnormality is tenuous and, in some cases, rather emotive.

To be able to make use of the sensitivity as a means of monitoring environmental pollution it is essential to recognise what is 'abnormal'. In every population of animals there are individuals that do not conform to the normal pattern and so there will be an 'acceptable' (expected) level of abnormalities. Thus in Wistar rats bred in captivity for laboratory studies about 3 per cent are abnormal; the incidence in the human population is roughly similar (J. Kirkwood, pers. comm.). Figures for stud cattle stocks are not available: it is unlikely that the owners would wish to admit that abnormalities occur in breeding lines. The important point is that abnormalities do and will occur in all species, and it is only when the incidence exceeds the norm that there is any cause for alarm.

WHAT IS NORMAL?

There is often a very poorly defined dividing line between 'normal' and 'abnormal', varying with the degree and diversity of variation in structures existing within any population of organisms. For example, although bilateral symmetry is one of the characteristics of all vertebrates it is apparent only at a very superficial level of expression. If you favour one side of your face when you grin you'll notice in the mirror that the facial skin folds are asymmetrical. At the skeletal level the differences between different sides of the body may be markedly different. The cranium and vertebral column of a frog superimposed upon a reference grid (Figure 62) demonstrates the extent of asymmetry in a 'normal' individual.

Because no two individuals are similar in every respect, abnormality may be viewed as variation exceeding the established limits for the population. For most parameters that can be quantified readily, normality is defined easily. For example, most frogs have four fingers and five toes. Hence three or five fingers or four or six toes would be considered abnormal.

Abnormalities expressed qualitatively are defined less readily. For example, if the angle of inclination of a vertebral process to the centrum varies, it may be difficult to specify the range above or below which an abnormality can be said to be exhibited. The architectural specification of the phenotype may not be defined well, and the degree of variation may therefore be extensive.

If follows that abnormalities generally have to be of a somewhat gross form to be recognised, but the situation is simplified a little by the fact that the abnormalities tend to be repetitive: appearing at high frequency in different samples.

THE NATURE OF ABNORMALITIES

Abnormalities of the skeletal system are named according to the condition and the structure that is affected. Hence the presence of supernumerary digits is termed 'polydactyly'

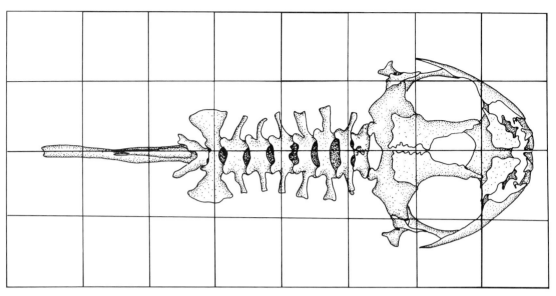

FIGURE 62 *Skull and vertebral column of a frog superimposed upon a grid, enabling comparison of the two halves and making it possible to observe minor deviation from symmetry.*

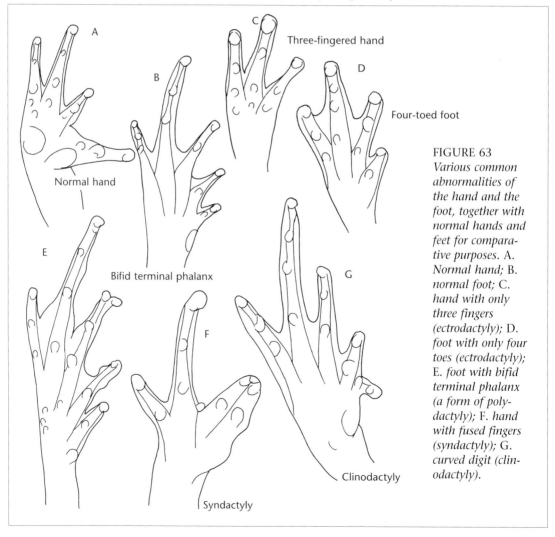

FIGURE 63
Various common abnormalities of the hand and the foot, together with normal hands and feet for comparative purposes. A. Normal hand; B. normal foot; C. hand with only three fingers (ectrodactyly); D. foot with only four toes (ectrodactyly); E. foot with bifid terminal phalanx (a form of polydactyly); F. hand with fused fingers (syndactyly); G. curved digit (clinodactyly).

and fusion of digits 'syndactyly'. The most common forms of anomaly are described below. The list is not exhaustive. For comparative purposes a normal hand and foot are shown in Figures 63A-B.

1 *Polydactyly*: Entire supernumerary digits occur almost without exception on the lateral or medial edges of customary digits. They occur much more frequently on the feet than on the hands.

Two variants attributed to constitute polydactyly have been reported in the literature. The first of these is the so-called 'weak polydactyly' ('Polydactylie faible') of Fischer (1973) characterised by a bilateral terminal phalanx (Figure 63E). The second condition described by Fischer (1977) is confined to the distal portion of the bone of the terminal phalanx, so creating a Y-shaped structure; but to judge from Fischer's X-ray of the example in *Rana*, this form of anomaly may not be visible from external view. Fischer regards this condition a morphological mimic of weak polydactyly. Hence such an interpretation requires an assumption that a digit may be duplicated at virtually any site along its length; one extreme being as a complete digit with an origin upon the distal head of a metatarsal or metacarpal; it may be a partial digit arising from a more proximal position upon a phalanx, or even as a division at a site upon the ultimate phalangeal bone.

2 *Syndactyly*: Syndactyly is the fusion of two or more digits of the hand (Figure 63F) or foot.

The fusion may be confined to soft tissues or may involve the uniting of skeletal components, including the phalanges and on occasions the metacarpals (or metatarsals) proximal to them as well. Gruneberg (1963) describes identical forms of syndactyly in mammals.

3 *Ectrodactyly* ('Oligodactyly'): Whereas syndactyly reduces the number of digits as a result of fusion, ectrodactyly is the loss of one or more digits without any evidence of the existence of fusion. Clearly the distinction may only be detected by radiographic or clearing and staining techniques. In the condition of oligosyndactyly, reported in

mice by Gruneberg (1956), the abnormality may be expressed either as syndactyly or as ectrodactyly.

Ectrodactyly may be partial (in which only a portion of the digit may be missing), or complete. If complete, the proximal metacarpal or metatarsal may be present—so that the affected digit is readily identifiable, or absent—so maintaining the symmetry of the hand and foot, and rendering identification of the affected digit much more difficult. (See the three-fingered hand and four-toed foot in Figures 63C-D).

4 *Clinodactyly*: By definition clinodactyly is a bent or curved digit. In practice the individual phalangeal bones are normal, but the terminal bone is inclined at an obtuse or acute angle to the penultimate bone (Figure 63G). However obtuse inclinations are exceptionally common post-mortem, with the result that often it is only when there is bilateral expression that the condition is recognisable in preserved specimens. In living frogs the condition can be detected quite readily.

5 *Hemimely*: Hemimely is the loss of all or part of the distal half of a limb (Figure 64A). Generally it is difficult to distinguish from amputation, and often identification hinges upon the absence of those features diagnostic of trauma, i.e. wound healing of superficial soft tissues and/or of bone.

6 *Ectromely*: Ectromely is the complete absence of a limb and hence, and in the case of hemimely, is an abnormality that conceivably could mimic trauma.

7 *Brachymely*: Whereas each member of a pair of limbs is normally identical in length, individuals are sometimes found in which one limb or a portion of a limb is smaller than the other (Figure 64B). This abnormality is termed brachymely. It may constitute negative allometric growth relative to the body, but it is indistinguishable from hyporegeneration following amputation of the growing limb or limb portion of a tadpole.

8 *Polymely*: Supernumerary limbs (polymely) attract more attention and comment than

PLATE 50 Limnodynastes dumerilii *with supernumerary legs. (Queensland Museum)*

any other form of abnormality, and Van Valen (1974) has assembled an extensive (but by no means exhaustive) list of published reports. Supernumerary hindlimbs are encountered much more frequently than forelimbs, and generally they are innervated and completely functional, as in the case of the individual of *Limnodynastes dumerilii* in Plate 50. Occasionally frogs are found with more than one supernumerary limb; Tyler (1976b, Plate 33) shows an Australian frog with two supernumerary forelimbs.

Supernumerary limbs can be complete, with all of the limb elements present, or reduced to the distal long bones, or even just the hand or foot. Some of the most bizarre forms of reduction are illustrated by Rostand (1958).

9 *Taumely*: Taumely is a gross disturbance of the plan of a limb, whereby a long bone is oriented 90° out of alignment (Figure 64C).

10 *Anophthalmy*: This is the absence of one or both eyes and has been observed in early stage embryos (where, if it is bilateral, it is obviously fatal). The absence of an eye in an adult could be attributed to trauma.

11 *Mandibular hypoplasia*: The failure of the mandible to grow at the same rate as the upper jaw results in a gap on the undersurface of the jaw. It appears to have been observed only in juvenile frogs suggesting that it interferes with the efficiency of the feeding process and so results in death.

12 *Kinky tail*: This condition is one of the major abnormalities in tadpoles and involves the presence of a deep bend at the base of the tail. If the tadpoles succeed in metamorphosing into frogs, the adults appear normal.

13 *Abbreviation of tail*: This abnormality is self explanatory. The tail may be only one-half of the normal length.

INJURIES AND REGENERATION

To be able to make any use of abnormality data the various causes have to be identified. But at first there has to be some way of distinguishing the abnormalities that are

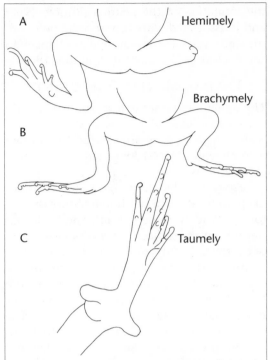

FIGURE 64 *Various common abnormalities of the hindlimb. A. Hemimely: where the distal portion of the limb is absent; B. Brachymely: where one limb is shorter than another in its proportions; C. Taumely: a gross disturbance where a limb element is oriented at right angles to the long axis of the limb.*

attributable to injury.

The frequency with which injury occurs in natural populations is uncertain. In France, Dubois (1979) placed the figure in a sample of more than 4000 *Rana esculenta* as high as 3 per cent. The extent of injury can be remarkable. Tyler, Davies and Walker (1985) reported injuries suffered by 17 *Neobatrachus aquilonius* which had fallen into an empty swimming pool on the Barkly Highway, Northern Territory, and which had then attempted to burrow into the concrete floor. The ends of the digits were abraded through to the bone, or phalangeal bones were missing and, in some cases, the side of the head had been shaved away through to the teeth. These severe injuries were self-inflicted. They testify to either a tolerance to, or lack of awareness of, pain and perhaps say something about the level of intelligence in these animals!

These horrific results of self-mutilation occurred in an artificial situation and, but for their retrieval by a collector, the frogs no doubt would have died. Nevertheless at least similar digital injuries could occur in a burrowing frog encountering a large flat rock during its passage through the soil.

It has long been known that tadpoles have the capacity to regenerate lost hind limbs, and that the degree of perfection of the regenerated limb is associated with the stage of development attained at the time of amputation. Thus complete regeneration of a limb will follow removal of the embryonic limb bud but a malformed limb will result if the long bones of the limb are well developed at the time of amputation. After metamorphosis has taken place regenerative capacity is lost. It is almost as though there is a competition between the processes of regeneration and wound-healing. Regeneration requires there to be a programme for replacement somehow coded within the stump. This attribute is lost progressively until wound-healing (which is oriented to maintaining the status quo) gains the upper hand, and effects a simple repair.

If partial regeneration produces an abnormal limb the question that must be addressed is whether any abnormalities observed in frogs in the field are attributable to injury suffered when they were tadpoles. From a consideration of the predators of tadpoles it is hard to imagine many situations in which the tadpole in such an encounter with one could escape losing only part of a leg. (Major predators such as dytiscid beetles and larvae, dragonfly larvae, water scorpions, etc. grab the tadpole, body and tail.) It is even harder to conceive of a bunch of inept predators all grabbing tadpoles by a limb, and all losing their prey—the ability of predators to be predators requires a high efficiency of predation or they would die of starvation. Nevertheless, because some of the abnormalities resulting from amputation or other artificially induced injury resemble those found in some frogs, injury must be considered. But if the incidence of abnormality in a population is high I think it is reasonable to exclude injury as the major causative factor.

IONISING RADIATION

Because the frog egg and developing tadpole can be observed directly, they are popular subjects for studies on the effects of ionising radiation on developmental processes. In addition they are considered the most radio-sensitive of aquatic animals (Emery & McShane, 1980). These studies have contributed substantially to a basic understanding of the effects of radiation upon higher animals.

Occasionally there are reports of malformed frogs being found in areas polluted with radioactive waste; reports, I might add, that are generally denied by the local authorities. One of the most celebrated was the discovery of grossly abnormal frogs in a canal carrying waste from a nuclear research institute in Amsterdam (Hillenius, 1959).

Hillenius observed that the abnormalities resembled those recorded at Trevignon, France by Rostand (1957) who had named the condition 'Anomaly P.'. Rostand was a distinguished academician and, whilst agreeing that the condition of the malformation at Amsterdam was 'Anomaly P.' he refuted the possibility of a cause associated with radio-activity. His argument hinged upon the absence of radioactive sources in the water at Trevignon (Rostand, 1957) and the apparent assumption that a manifestation could not be produced by more than one causal agent.

Henle (1981) reported a curious experience at a quarry about 25 km from Stuttgart in Germany where discovery of a large number of abnormal toads *(Bufo viridis)* was followed by what Henle clearly considered to be a cover-up operation by the owners of the quarry and local authorities. These experiences do not create an atmosphere in which radioactivity as an inducer of abnormal growth can be elucidated. It is certainly necessary to assert that it is indeed a source of abnormal growth. For example, Nishimura (1967) produced a variety of deformities in frogs reared from rainwater contaminated with radioactive dust, obtained following nuclear explosion tests on Bikini Atoll in March 1954 and Nevada in April 1955.

Whole body irradiation is known to produce genetic damage and this varies both in incidence and expression in subsequent generations (Blair, 1960).

Sensitivity to radioactive substances appears to decrease with the age of the tadpole. The most sensitive stage is that prior to the formation of the neurula (Blinov, 1962) and is associated with very specific points in the cell cycles corresponding to changes in the nuclear DNA content (Hamilton, 1969).

ENVIRONMENTAL POLLUTANTS

The aquatic environment is polluted with a very wide range of chemical substances of which the following are the most common: heavy metals, insecticides, herbicides, fungicides, aflotoxins.

Frogs are obviously significant creatures likely to be affected but the literature on this topic is widely dispersed and the topic has never been the subject of a review either here or overseas. An 'Inventory of data on contaminants in aquatic organisms' produced by the Food and Agriculture Organisation of the United Nations runs to 397 pages, but does not include a single reference to frogs. The most constructive approach seems to be to outline the current state of knowledge and provide key references.

DDT was the first of the organochlorine insecticides. It was first manufactured in 1874 but its value as an insecticide was not appreciated until the 1940s. It has been used very widely in Australia for both the control of vectors of human diseases, and for the control of insect pests of crops. Connel (1974) estimated the annual use of DDT in Australia at 900 tonnes.

The benefits of DDT are indisputable but it is apparent that in some areas applications were very high, and it was a result of the expression of concern about its impact upon non-target organisms that led to more stringent controls being applied to it and other compounds. The high levels of DDT application had to be 'sold' to the general public as beneficial. At Kununurra, Western Australia, a tourist brochure drew the attention of visitors to DDT application by referring to the 'majestic clouds' following the crop-dusters.

The literature on the effects of DDT on frogs is fairly extensive. An example is the studies of its effects upon the spawn and tadpoles of the European frog *Rana temporaria* (Cooke, 1970, 1972, 1973a, 1973b). Tadpoles became hyperactive and developed abnormalities of the snout. Hyperactivity occurred in individuals with tissue concentrations of as little as 0.1 parts per million (ppm), and had a major effect upon development: the hyperactivity causing the lower mandible to strike the inner surface above the upper jaw with sufficient force to cause the eventual detachment of the upper jaw. Although some individuals recover, metamorphosing individuals commonly had blunt snouts and deformed brains (Osborn *et al.*, 1981).

DDT accumulates in fat bodies and there is concern for those animals at the top of food chains involving a variety of creatures exposed to it. Jaskoski and Kinders (1974) found DDT in the fat bodies of three North American frog species six years after the final DDT spraying of the areas which they inhabited. Fleet, Clark and Plapp (1972) examined residues in snakes that feed upon frogs and found the DDT metabolite DDE in fat at levels of up to 1009.4 ppm. This maximum occurred in the Cottonmouth snake. In Australia, Best (1973) reported a variety of organochlorine pesticide residues, including DDT, in native fauna, and demonstrated that the highest amounts occurred in animals at the top of the food chain from developed areas. More recently Birks and Olsen (1987), reporting a 1972 study, examined pesticide concentrations in diverse South Australian vertebrates (including *Limnodynastes tasmaniensis*) without noticing any hierarchy in the food chain.

Reports of the effects of DDT upon frogs have involved a wide variety of morphological and physiological studies, including the popular area of regeneration. Weis (1975) noted marked inhibition of regenerative capacity of tadpole tails even at exposures to 25 parts per billion (25 ppb). Such sensitivity is staggering and indicates just how minor levels of pollution can have a potential effect upon frogs. Sanders (1970) compared the effects of a variety of pesticides upon two North American species and noted that the sensitivity of tadpoles to DDT increased with age over a period of one to seven weeks. The results of this and other studies on insecticides are recorded in Table 15.

TABLE FIFTEEN: Studies on the effects of insecticides upon frogs

COMPOUND	TEST SPECIES	TOXICITY DATA	ABNOR- MALITY DATA	OTHER	REFERENCE
Aldrin	*Acris crepitans*	+	—	—	Ferguson & Gilbert (1967)
"	*A. gryllus*	+	—	—	"
"	*Bufo fowleri*	+	—	—	"
"	*B. bufo*	+	—	—	Ludemann & Neumann (1961)
"	*B. woodhousii*	+	—	—	Sanders (1970)
"	*Pseudacris triseriata*	+	—	—	"
"	*Rana cyanophlyctis*	—	—	—	Rane & Mathur (1978)
"	*R. pipiens*	+	—	—	Kaplan & Overpeck (1964)
Altosid (2R–515)	*Bufo bufo*	+	—	—	Paulov (1976)
Benzene hexachloride	*B. woodhousii*	+	—	—	Sanders (1970)
"	*Pseudacris triseriata*	+	—	—	"
"	*Rana pipiens*	+	—	—	Kaplan & Overpeck (1964)

COMPOUND	TEST SPECIES	TOXICITY DATA	ABNOR-MALITY DATA	OTHER	REFERENCE
Carbophenothion	*Pseudacris triseriata*	+	—	—	Sanders (1970)
"	*Bufo woodhousii*	+	—	—	"
Chlordan	*B. bufo*	+	—	—	Ludemann & Neumann (1961)
Chlordane	*Rana pipiens*	+	—	—	Kaplan & Overpeck (1964)
Chlorthion	*Bufo bufo*	+	—	—	Neumann (1967)
DDT	*Acris crepitans*	+	—	—	Ferguson & Gilbert (1967)
"	*A. gryllus*	+	—	—	"
"	*Bufo bufo*	+	—	—	Marchal-Segault (1976)
	"	+	+	hyperactivity	Cooke (1972)
	"	+	—	—	Ludemann & Neumann (1962)
"	*B. fowleri*	+	—	—	Ferguson & Gilbert (1967)
"	*B. woodhousii*	+	—	—	Sanders (1970)
"	*Limnodynastes tasmaniensis*	—	—	residues	Birks & Olsen (1987)
"	*Pseudacris triseriata*	+	—	—	Sanders (1970) (1970)
"	*Rana catesbeiana*	—	—	regeneration	Weis (1975)
	"	—	—	residues	Jaskoski & Kinders (1974)
	"	—	—	"	Meeks (1968)
	R. clamitans	—	—	—	"
"	*R. pipiens*	—	—	regeneration	Weis (1975)
	"	—	—	residues	Jaskoski & Kinders (1974)
	"	—	—	"	Meeks (1968)
"	*Rana sylvatica*	—	—	residues	Licht (1976)
"	*R. temporaria*	+	+	hyperactivity	Cooke (1973a)
	"	—	+	"	Osborn *et al.* (1981)
	"	+	+	"	Cooke (1972)
	"	+	—	—	Harris *et al.* (1979)
	"	—	—	residues	Cooke (1974), (1979)
DFP	*Rana ridibunda*	—	—	—	Edery & Schatzberg-Porath (1960)

COMPOUND	TEST SPECIES	TOXICITY DATA	ABNOR-MALITY DATA	OTHER	REFERENCE
Diazinon & Neumann	*Bufo bufo*	—	—	—	Ludeman (1962)
Dieldrin	"	+	+	hyperactivity	Cooke (1972)
	"	—	—	—	Ludemann & Neumann (1962)
"	*B. woodhousii*	+	—	—	Sanders (1970)
"	*Limnodynastes tasmaniensis*	+	+	—	Brooks (1981)
	"	—	—	residues	Birks & Olsen (1987)
"	*Pseudacris triseriata*	+	—	—	"
"	*Rana pipiens*	+	—	—	Kaplan & Overpeck(1964)
"	*R. temporaria*	—	+	hyperactivity	Cooke (1972)
	Acris crepitans	+	—	—	Ferguson & Gilbert (1967)
"	*A. gryllus*	+	—	—	"
"	*Bufo fowleri*	+	—	—	"
Dimefox	*B. viridis*	+	—	—	Edery & Schatzberg-Porath (1960)
"	*Rana ridibunda*	+	—	—	"
Diphanos	*R. chensinensis*	+	+	—	Hattori (1974)
Dipterex	*Bufo bufo*	+	—	—	Ludemann & Neumann (1962)
Endrin	*Acris crepitans*	+	—	—	Ferguson & Gilbert (1967)
"	*A. gryllus*	+	—	—	"
"	*Bufo bufo*	+	—	—	Ludemann & Neumann (1962)
"	*B. fowleri*	+	—	—	Ferguson & Gilbert (1967)
"	*B. woodhousii*	+	—	—	Sanders (1970)
"	*Pseudacris triseriata*	+	—	—	"
"	*Rana cyanophlyctis*	—	—	histology	Mathur & Ram (1978)
"	*R. temporaria*	+	—	—	Wohlgemuth (1977)
	"	+	—	—	Wohlgemuth & Trnkova (1979)
"	*R. esculenta*	+	—	—	"

COMPOUND	TEST SPECIES	TOXICITY DATA	ABNOR-MALITY DATA	OTHER	REFERENCE
"	*R. sphenocephala*	+	—	residues	Hall & Swineford (1980)
"	*R. pipiens*	+	—	—	Kaplan & Overpeck (1964)
Eulan New WA	*R. temporaria*	+	+	residues	Osborn & French (1981)
Fenthion	*Bufo bufo*	+	—	—	Marchal-Segault (1976)
"	*Rana catesbeiana*	+	—	—	Hall & Kolbe (1980)
Folithion	*R. tigerina*	+	—	—	Mohanty, Hejmadi & Dutton (1982)
Foschlorine	*R. temporaria*	+	—	—	Ranke & Rybicka (1972)
Gesagard 50	"	+	—	—	Jordan *et al.* (1977)
Guthion	*Bufo woodhousii*	+	—	—	Sanders (1970)
"	*Pseudacris triseriata*	+	—	—	"
Heptachlor	*Bufo bufo*	+	—	—	Ludemann & Neumann (1962)
"	*B. woodhousii*	+	—	—	Sanders (1970)
Karbatox 75	*Rana temporaria*	+	+	hyperactivity	Rzchak et al. (1977)
"	*Xenopus laevis*	+	+	"	"
Lindane	*Bufo bufo*	+	—	—	Ludemann & Neumann (1962)
"	*B. woodhousii*	+	—	—	Sanders (1970)
"	*Pseudacris triseriata*	+	—	—	"
Malathion	*Bufo bufo*	+	—	—	Ludemann & Neumann (1962)
"	*B. woodhousii*	+	—	—	Sanders (1970)
"	*Pseudacris triseriata*	+	—	—	"
"	*Rana catesbeiana*	+	—	—	Hall & Kolbe (1980)
"	*R. temporaria*	+	—	—	Ranke & Rybicki (1972)

COMPOUND	TEST SPECIES	TOXICITY DATA	ABNOR-MALITY DATA	OTHER	REFERENCE
"	*R. tigerina*	+	—	—	Mohanty-Hejmadi & Dutton (1981)
Metacid	"	+	—	—	"
Metasystox	*Bufo bufo*	+	—	—	Ludemann & Neumann(1962)
"	*Rana tigerina*	+	—	—	Mohanty-Hejmadi & Dutton (1981)
Methoxychlor	*Bufo americanus*	—	—	residues	Hall & Swineford (1979)
"	*B. woodhousii*	+	—	—	Sanders (1970)
"	*Pseudacris triseriata*	+	—	—	"
"	*Rana pipiens*	+	—	—	Kaplan & Overpeck (1964)
Neled	*Bufo woodhousii*	+	—	—	Sanders (1970)
"	*Pseudacris triseriata*	+	—		"
Oxamyl	*Rana temporaria*	+	+	—	Cooke (1981)
Paraoxon	*Bufo viridis*	+	—	—	Edery & Schatzberg-Porath (1960)
"	*Rana ridibunda*	+	—	—	"
Parathion	*Bufo bufo*	+	—	—	Ludemann & Neumann (1962)
"	*B. viridis*	+	—	—	Edery & Schatzberg-Porath (1960)
"	*B. woodhousii*	+	—	—	Sanders (1970)
"	*Pseudacris triseriata*	+	—	—	"
"	*Rana ridibunda*	+	—	—	Edery & Schatzberg-Porath (1960)
"	*R. catesbeiana*	+	—	—	Hall & Kolbe (1980)
Permethrin	*Rana catesbeiana*	+	—	—	Jolly *et al.* (1978)

COMPOUND	TEST SPECIES	TOXICITY DATA	ABNOR-MALITY DATA	OTHER	REFERENCE
Piperonyl butoxide	*Bufo woodhousii*	+	—	—	Sanders (1970)
"	*Pseudacris triseriata*	+	—	—	"
Rogor	*Rana tigerina*	+	—	—	Mohanty-Hejmadi & Dutton (1981)
"	"	+	—	—	Dutton, Mohanty-Hejmadi (1978)
Sevin	*Bufo bufo*	+	—	—	Marchal-Segault (1976)
Systox	"	+	—	—	Ludemann & Neumann (1962)
TDE	*Bufo woodhousii*	+	—	—	Sanders (1970)
"	*Pseudacris triseriata*	+	—	—	"
TEPP	*Bufo viridis*	+	—	—	Edery & Schatzberg-Porath (1960)
"	*Rana ridibunda*	+	—	—	"
2,3,7,8— Tetra chlorodibenzo—p—dioxin	*R. catesbeiana*	+	—	—	Beatty *et al.* (1976)
Thiodam	*Bufo bufo*	+	—	—	Ludemann & Neumann (1962)
Toxaphene	*Acris crepitans*	+	—	—	Ferguson & Gilbert (1967)
"	*A. gryllus*	+	—	—	"
"	*Bufo fowleri*	+	—	—	"
"	*B. bufo*	+	—	—	Ludemann & Neuman (1962)
"	*B. woodhousii*	+	—	—	Sanders (1970)
Toxaphene	*Pseudacris triseriata*	+	—	—	"
"	*Rana pipiens*	+	—	—	Kaplan & Overpeck (1964)
	R. sphenocephala	+	—	residues	Hall & Swineford (1980)
Tritox—30	*R. temporaria*	+	—	—	Wojcik, Ranke & Rybicka (1971)
	Xenopus laevis	+	—	—	"

Herbicides and fungicides have attracted less interest than insecticides, and the majority of papers published have been concerned with establishing survival of frogs and tadpoles to their exposure rather than eliciting whether they can produce abnormalities (Table 16).

TABLE SIXTEEN: Studies of the effects of herbicides and fungicides upon frogs

COMPOUND	TEST SPECIES	TOXICITY DATA	ABNOR-MALITY OBSERVED	REFERENCE
2,4,5-T amine	*Adelotus brevis*	+	—	Johnson (1976)
"	*Bufo marinus*	+	—	"
"	*Limnodynastes peronii*	+	—	"
Ansar 529 E IC	*Scaphiopus couchi*	+	—	Judd (1977)
Atrazine	*Rana temporaria*	+	+	Hazelwood (1970)
"	*Bufo vulgaris*	+	—	Costantini & Panella (1974)
Chloranil	*Xenopus laevis*	+	+	Anderson & Prahlad (1976)
6-chlora-2 picolinic acid	*Bufo woodhousii*	+	—	Sanders (1970)
"	*Pseudacris triseriata*	+	—	"
Cynatryn	*Rana temporaria*	+	—	Scorgie & Cooke (1979)
2,4-D amine	*Adelotus brevis*	+	—	Johnson (1976)
"	*Bufo marinus*	+	—	"
"	*Limnodynastes peronii*	+	—	"
DEF	*Pseudacris triseriata*	+	—	Sanders (1970)
"	*Bufo woodhousii*	+	—	"
Defenuron	*Rana temporaria*	+	—	Paulov (1977b)
Dicamba	*Adelotus brevis*	+	—	Johnson (1976)
"	*Bufo marinus*	+	—	"
"	*Limnodynastes peronii*	+	—	"
Dichlone	*Xenopus laevis*	+	+	Anderson & Prahlad (1976)
Diquat	" "	+	+	"
"	*Rana pipiens*	+	+	Bimber & Mitchell (1978)
"	" "	+	—	Dial & Bauer Dial (1987)
Dosanex	*Bufo vulgaris*	+	—	Costantini & Andreoli (1972)
"	" "	+	—	Costantini & Panella (1974)

COMPOUND	TEST SPECIES	TOXICITY DATA	ABNOR- MALITY OBSERVED	REFERENCE
2,2-DPA	*Adelotus brevis*	+	—	Johnson (1976)
"	*Bufo marinus*	+	—	"
"	*Limnodynastes peronii*	+	—	"
DSMA	*Adelotus brevis*	+	—	"
"	*Bufo marinus*	+	—	"
"	*Limnodynastes peronii*	+	—	"
Fenac	*Pseudacris triseriata*	+	—	Sanders (1970)
"	*Bufo woodhousii*	+	—	"
Fenoprop	*Adelotus brevis*	+	—	Johnson (1976)
"	*Bufo marinus*	+	—	"
"	*Limnodynastes peronii*	+	—	"
"	*Litoria ewingi*	+	—	"
Gramoxone S (paraquat dichloride)	*Rana temporaria*	+	—	Paulov (1977a)
Hydrothol 191	*Bufo woodhousii*	+	—	Sanders (1970)
	Pseudacris triseriata	+	—	"
Igran	*Bufo vulgaris*	+	—	Costantini & Panella (1974)
Lasso	" "	+	–	"
Maneb	*Xenopus laevis*	+	+	Bancroft & Prahlad (1973)
Molinate	*Pseudacris triseriata*	+	—	Sanders (1970)
"	*Bufo woodhousii*	+	—	"
Monosodium methylarsonate	*Scaphiopus couchi*	+	—	Judd (1977)
MSMA	*Adelotus brevis*	+	—	Johnson (1976)
"	*Bufo marinus*	+	—	"
"	*Limnodynastes peronii*	+	—	"
Nabam	*Xenopus laevis*	+	+	Anderson & Prahlad (1976)
	" "	+	+	Bancroft & Prahlad (1973)
Na TA	*Bufo vulgaris*	+	—	Costantini & Panella (1974)
Paraquat	*B. woodhousii*	+	—	Sanders (1970)
"	*Pseudaris triseriata*	+	—	"

COMPOUND	TEST SPECIES	TOXICITY DATA	ABNOR- MALITY OBSERVED	REFERENCE
"	*Rana pipiens*	+	+	Dial & Bauer (1984)
"	" "	+	—	Dial & Bauer Dial (1987)
"	*Adelotus brevis*	+	—	Johnson (1976)
"	*Bufo marinus*	+	—	"
"	*Limnodynastes peronii*	+	—	"
Picloram	*Adelotus brevis*	+	—	"
"	*Bufo marinus*	+	—	"
"	*Limnodynastes peronii*	+	—	"
Pyramin	*Bufo vulgaris*	+	—	Costantini & Andreoli (1972)
"	" "	+	—	Costantini & Panella (1974)
Silvex	*Pseudacris triseriata*	+	—	Sanders (1970)
"	*Bufo woodhousii*	+	—	"
Sodium arsenite	*Adelotus brevis*	+	—	Johnson (1976)
"	*Bufo marinus*	+	—	"
"	*Limnodynastes peronii*	+	—	"
Tribunil	*Bufo vulgaris*	+	—	Costantini & Andreoli (1972)
"	" "	+	—	Costantini & Panella (1974)
Trifluranil	*Pseudacris triseriata*	+	—	Sanders (1970)
"	*Bufo woodhousii*	+	—	"
Weeder 64 (amine salt of 2,4-D)	" "	+	—	()
"	*Pseudacris triseriata*	+	—	"

Amongst the trace elements, selenium has been shown to produce severe developmental abnormalities at concentrations of 2 ppm (Browne and Dumont, 1979). The motivation for such studies occurred on that occasion because of the known toxicity of the substance, an awareness of the likelihood of selenium becoming more widespread in the environment, and the need to develop a new monitoring system. Work on the fungal toxins Aflotoxin B, and Aflotoxin G (Gabor *et al.*, 1973; Puscariu *et al.*, 1973) appears to have been similarly motivated.

Despite their relevance to environmental studies, relatively few papers report the effect of metals but a broad selection is presented in Table 17. In Finland, Posanen and Koskela (1976) have demonstrated seasonal variation in the levels in various body tissues and organs of zinc, calcium and magnesium.

TABLE SEVENTEEN: **Presence in tissues and effects of heavy and transitional metals upon frogs**

METAL	TEST SPECIES	TOXICITY	ABNOR-MALITY	REFERENCE
Arsenic	*Bombina variegata*	—	—	Byrne *et al.* (1975)
"	*Bufo bufo*	—	—	"
"	*Rana catesbeiana*	+	—	Birge & Just (1973)
"	*R. esculenta*	—	—	Byrne *et al.* (1975)
"	*R. pipiens*	+	—	Birge & Just (1973)
	R. temporaria	—	—	Bryne *et al.* (1975)
Beryllium	*R. pipiens*	+	—	Dilling & Healey (1925)
Cadmium	*Bufo arenarum*	+	+	Perez *et al.* (1985)
"	*Gastrophryne carolinensis*	+	+	Birge *et al.* (1977)
Chromium	*Rana esculenta*	—	—	Bando (1976)
"	*R. tigerina*	+	+	Abbassi & Soni (1984)
Copper	*Alytes obstetricans*	+	—	Deschiens *et al.* (1965)
"	*Bombina variegata*	—	—	Byrne *et al.* (1975)
"	*Bufo bufo*	—	—	"
"	*Rana esculenta*	—	—	"
"	" "	+	—	Bando (1976)
"	*R. pipiens*	+	—	Dilling & Healey (1925)
"	"	+	—	Kaplan & Yoh (1961)
"	"	+	—	Lande & Guttman (1973)
"	*R. temporaria*	—	—	Byrne *et al.* (1975)
"	*Xenopus laevis*	+	—	Fingal & Kaplan (1963)
Lead	*Rana catesbeiana*	+	—	Birge & Just (1973)
"	*R. esculenta*	—	—	Bando (1976)
"	*R. pipiens*	+	—	Dilling & Healey (1925)
"	"	+	—	Birge & Just (1973)
Lithium	*Bufo arenarum*	—	+	Bustuoabad *et al.* (1977)
"	*Rana catesbeiana*	+	—	Birge & Just (1973)
"	*R. fusca*	+	+	Pasteels (1942, 1945)
	R. pipiens	+	—	Birge & Just (1973)
"	*Xenopus laevis*	—	+	Backstrom (1953)
Manganese	*Rana esculenta*	—	—	Bando (1976)
Mercury	*Bombina variegata*	—	—	Byrne *et al.* (1975)
"	*Bufo bufo*	—	—	"
"	*Gastrophryne carolinensis*	+	+	Birge *et al.* (1977)

METAL	TEST SPECIES	TOXICITY	ABNOR-MALITY	REFERENCE
"	*Microhyla ornata*	+	+	Ghate & Mulherkar (1980)
"	*Rana catesbeiana*	+	+	Birge & Just (1973)
"	*R. esculenta*	—	—	Byrne *et al.* (1975)
"	*R. pipiens*	+	+	Birge & Just (1973)
		+	+	Dial (1976)
"	*R. temporaria*	—	—	Byrne *et al.* (1975)
"	*Xenopus laevis*	+	+	Schewing & Boverio (1979)
Thallium	*Rana pipiens*	+	—	Dilling & Healey (1925)
Thorium	"	+	—	''
Zinc	*Bombina variegata*	—	—	Byrne *et al.* (1975)
"	*Bufo bufo*	—	—	"
"	*Gastrophryne carolinensis*	+	—	Birge *et al.* (1977)
"	*Rana catesbeiana*	+	—	Birge & Just (1973)
"	*R. esculenta*	—	—	Byrne *et al.* (1975)
"	*R. pipiens*	+	—	Dilling & Healey (1925)
"	"	+	—	Birge & Just (1973)
"	*R. temporaria*	—	—	Byrne *et al.* (1975)
"	"			Pasanan & Koskela (1976)

THE JABIRU EXPERIENCE

The Fox Report on the development of uranium mining in the Kakadu National Park in the Northern Territory, recommended that monitoring systems be established capable of demonstrating any environmental changes. The Office of the Supervising Scientist for the Alligator Rivers Region (OSS) was set up by the Australian government to coordinate and promote this work.

In 1977, prior to the development of the mining phase, Margaret Davies and I visited the area and collected spawn from two sites: one at Radon Creek (named for its high radon content) and the other from a gravel pit far away from it. The samples were returned to Adelaide, reared into frogs and were then examined. The Radon Creek sample exhibited a high level of abnormalities of the limbs, and the question arose as to whether radon was responsible. As a result a research program funded by the OSS was commenced to examine the incidence of skeletal abnormalities in the area. This was an important step in the pre-mining phase because it would provide a base level against which any surveys during mining could be compared. If mining changed conditions sufficiently to affect the incidence of abnormalities, we could use the frogs as a new environmental monitoring organism.

The first step in our programme was simply to establish which frogs occurred in the area. Of our final total of 22 species, six were new to science, reflecting the poor overall knowledge of the frog fauna of the Top End of the Northern Territory. We then had to decide which species would be most significant in accumulating radionucleides, and this involved working out how much time was spent feeding in or near water. Such species would be likely to colonise the vast tailings dam and retention ponds yet to be constructed.

One would have thought that water at Jabiru, so far from major centres, would at least be free of herbicide and insecticide residues. Unfortunately this was not the case: helicopter pads had been sprayed with 2,4,5, -T and 2,4D to inhibit weed growth, roadside marker poles had been treated to prevent termite damage, water level monitoring stations had been treated with herbicides to prevent macrophyte growth. It was obviously necessary to establish whether these herbicides and pesticides could be responsible for any of the abnormalities observed in the area.

The problem is that you need to know in advance the kinds of herbicides and insecticides used. It is not possible to pose the question 'Are any herbicides or insecticides present?—the chemist has to be asked to test for the presence of specific compounds. Tests for some of the more common pesticides and herbicides revealed levels of no more than 0.5 ppm which is approaching the limits of analysis and insufficient to induce abnormalities.

We also needed data on abnormality levels in frogs not exposed to pollution of any kind. These data were assembled from collections made in several countries (Table 18) and revealed levels of 0–3.9 per cent but with only one sample over 2 per cent.

TABLE EIGHTEEN: Incidence of abnormalities of limbs in control samples of frogs

SOURCE	GENUS AND SPECIES	NORMAL	ABNORMAL	% ABNORMAL
USA	*Scaphiopus bombifrons*	520	6	1.14
Ecuador	hylids—4 species	202	3	1.46
"	*Atelopus*—3 species	188	6	3.09
Costa Rica	*Eleutherodactus rugulosus*	92	1	1.08
Guatemala	*Agalychnis moreletti*	60	0	0
North Korea	*Bombina orientalis*	50	1	1.96
New Guinea	*Litoria*—3 species	439	2	0.45
"	*Nyctimystes kubori*	177	0	0
"	*Rana grisea*	59	1	1.67

TABLE NINETEEN: Incidence of limb injuries among frogs at Jabiru, Northern Territory

SPECIES	TOTAL SAMPLE	NUMBER INJURED	% INJURY
Cyclorana australis	939	5	0.53
C. longipes	522	1	0.19
Limnodynastes ornatus	458	2	0.44
Litoria dahlii	101	1	0.99
L. pallida	132	1	0.76
L. rubella	479	3	0.63
Uperoleia inundata	111	1	0.90

TABLE TWENTY: Comparison of incidence of abnormalities in juvenile and adult frogs

SPECIES	TOTAL	ADULTS ABNORMAL	% ABNORMAL	TOTAL	JUVENILES ABNORMAL	% ABNORMAL
Cyclorana australis	53	0	0	886	68	7.7
C. longipes	61	1	1.6	459	37	8.1
Limnodynastes ornatus	52	6	11.5	415	55	13.25

Similarly we examined the incidence of injury in 2742 specimens. Those reported for each different species was less than 1 per cent and ranged 0.19–0.99 per cent (Table 19). Although many of the abnormalities to the limbs appeared, in our judgment, extremely minor, it seemed that some were sufficient to cause the death of the individual. This assumption was based on comparison of the incidence of abnormalities in adults compared with juveniles from one locality (Table 20) demonstrating substantially higher abnormalities in juveniles.

By far the most commonly occurring abnormality was ectrodactyly (Table 21). The largest series was taken in a small body of water derived from a radioactive source and trapped within the walls of Retention Pond No. 2 at Jabiru prior to its filling. Spawning of *Cyclorana australis* and *C. Iongipes* took place there in early November 1979. The young completed metamorphosis between 3 and 14 of December. Fifty-eight were normal and 37 exhibited ectrodactyly (38.95 per cent). On 15 December there was heavy rain diluting the water considerably and the species again bred there. Of the frogs reared from the second spawning 300 were normal and only five exhibited ectrodactyly (1.6 per cent).

TABLE TWENTY-ONE: Occurrence of abnormalities in various species of frogs at Jabiru, Northern Territory.

	No. abnormal individuals	% abnormal individuals	Polydactyly	Syndactyly	Ectrodactyly	Clinodactyly	Hemimely	Ectromely	Anophthalmy	Taumely	Brachymely	Mandibular Hypoplasia	Othe (minor)	Other (major)	Total abnormalities
Crinia bilingua	0	0													0
Cyclorana australis	64	7.0	1		54	1	1		1		6			1	65
C. longipes	39	8.4		2	25	4				8			3	2	44
Limnodynastes convexicusculus	0	0													0
L. ornatus	60	15.7	1	32	26	1				1			10	1	72
Litoria bicolor	6	9.8		1	5										6
L. caerulea	5	16.1			1		2					1		1	5
L. coplandi	0	0													0
L. dahlii	13	12.9	1		7	2	1								13
L. inermis	7	8.0			5	2					2		2		9
L. meiriana	3	4.0			2			1			1		1		5
L. microbelos	0	0													0
L. nasuta	0	0													0
L. pallida	5	4.5			3	3									6
L. personata	0	0													0
L. rothii	1	3.8			1										1
L. rubella	36	8.6		2	19						1	15	2	2	41
L. tornieri	6	7.3			3	1							3		7
L. wotjulumensis	0	0													0
Megistolotis lignarius	0	0													0
Notaden melanoscaphus	7	5.4			3										3
Sphenophryne adelphe	0	0													0
Uperoleia arenicola	0	0													0
U. inundata	9	12.6			9	3							2		14
	261		2	38	163	17	4	1	1	9	10	16	23	7	291

The Jabiru studies provided a base line of the levels of abnormality occurring prior to major disturbance. There were no previous studies involving the use of frogs as environmental monitoring organisms and so problems encountered required novel solutions. However its value lies in providing a detailed statement with which comparisons can be made in the future.

ABNORMALITIES AND EVOLUTION

It has been noted by Grosch and Hopwood (1979) that some developmental abnormalities, that occurred following irradiation of larval and pupal stages of insects, mimic known mutational changes but are not inherited. Amongst Australian frogs some of the structural changes that have accompanied the divergence of species or genera mimic adaptive structures very closely. For example, at Paralana Springs in the northern Flinders Ranges there is a relatively high level of radon discharge (2000 p.Ci/litre). The froglet *Crinia riparia* breeding at the site exhibits a high level of abnormalities to the hands (35 per cent) including the development of flanges to the lateral surface of the first three fingers. These flanges appear identical to the sex-linked structures acquired by female limnodynastine frogs (including *Limnodynastes),* and which improve the efficiency of paddling during the construction of foam nests. It remains to be seen whether the structures in *C. riparia* are inherited.

Van Valen (1974) proposed that the high incidence of supernumerary structures in frogs could result in the evolution of higher taxa. Equally plausible is the incidence of loss of digital bones characterising genera such as *Arenophryne.* Both the acquisition of supernumerary structures and the loss of others assume adaptive benefit in some populations.

CHAPTER FOURTEEN

DECLINING FROG POPULATIONS

GLOBAL CONSIDERATIONS

In a historical sense frogs are among the senior citizens of the existing vertebrate animals. No doubt their fortunes have ebbed and waned from time to time, but they have had the capacity to persist and flourish despite climatic changes, mountain building, marine incursions of their environment, predation and competition.

Humans have shared this earth with frogs for an infinitesimal portion of the frogs' success story. It is actually equivalent to about 0.05 per cent of the frog ancestral span. Unfortunately, humans rarely live in harmony with animals, and frogs are no exception. Certainly there are attempts at coexistence, such as not objecting to a frog sharing the bathroom, and recent efforts in Europe to construct tunnels to permit migrating toads to pass under highways; but these actions are only of localised significance. They are events that make us feel good because we can cite them as tangible expressions of our concern to enable frogs to survive.

The reality of human impact is that we have created a situation in which the short- to medium-term existence of frogs on this earth does not look at all good. But probably the only way in which we will involve ourselves in halting, or reducing the process of decline and extinction, will be because we perceive human benefit from doing so.

Exploitation of the world to modify it for human life is an intrinsic part of our survival. Certainly in recent years there have been efforts to halt the process of deforestation, because we have recognised the impact upon the nature of the world's atmosphere and the threat posed to the human race. Frogs and other forest dwellers will benefit, but few have heeded their plight.

Insofar as habitat destruction brought about by other human activity is concerned, even the most biased conservationist should accept that some of it is inevitable. The economics of feeding, housing and clothing is such that a level of degradation just cannot be avoided.

Historically it would therefore be expected that the geographic range of frogs has resulted in their extinction in areas where there have been particularly radical modifications of the environment, such as within inner cities, or in areas of intense industrial development. Similarly reductions in abundance would be anticipated through urban consolidation and other less radical changes. Unfortunately frogs are also declining in some areas where as yet no obvious cause can be detected. Where have all the frogs gone?

In September 1989 the First World Congress of Herpetology was held in Canterbury, Kent. A total of 1000 delegates from 75 countries exchanged information on everything from frog evolution to the state of frogs today. A number of delegates from all continents presented the extremely disturbing news that numerous species were in serious decline, and that significant decline appeared to have commenced everywhere between 1978 and 1983. The synchrony of the declines seemed particularly perplexing.

Acting upon the initiative of David B. Wake of the University of California at Berkeley, the Board of Biology of the National Academy of Sciences urgently arranged a meeting of scientists which was held at Irvine, California in February of the following year. In addition to herpetologists they invited a few experts in climatic studies, water chemistry, animal diseases, population analysis and other fields considered potentially relevant.

The first question was to establish where there *was* a genuine and significant decline. The second was whether there was any evidence of a common causal factor. Responding to the issue of the decline one delegate after another substantiated the trend reported at the World Congress of Herpetology. Any caution they may have felt in expressing that concern did not extend to their comments to reporters from prestigious publications such as the *New York Times* and *Science*.

The dilemma faced was that frog populations have huge, and often rapid cyclic swings between being abundant and scarce. Sometimes the swings can be anticipated—for example, when a moist area ideal for frogs becomes an arid one in which survival of any is a miracle. But swings also seem to occur as a normal phenomenon. This has been demonstrated well at the Savannah River Ecology Laboratory in south-eastern California by Laurie Vitt. He was able to document wild fluctuations in the number and species of frogs whose movements were monitored by collecting them daily from specially excavated pits into which they had fallen. If the numbers decline over one period, is it not possible that they will increase later?

On a broader scale David Bradford of the University College of Los Angeles reported remarkable declines in frogs in the high mountains of the Sierra Nevada of California. Of the four species occurring at altitudes of 8000–12 000 feet (approximately 2400–3600 m) above sea-level, one, *Rana muscosa*, had been abundant at 38 lakes that he surveyed in 1977–80. This species inhabits streams and is active during the brief summer when it can be found basking in the sun. When Bradford returned to the lakes in 1989 he found that it had disappeared from 37 of them. In the same area the Yosemite toad had faced a similar decline, and the last female was seen there in 1986.

David Wake observed that salamanders had disappeared in the same area, and that he had observed dramatic declines in Costa Rica.

Evidence assembled in Europe and Scandinavia demonstrates massive local extinction over vast tracts of land, but whether a significant component of that decline has taken place within the last decade or so, does not seem to have been analysed. Data from Africa, Asia, Russia and other Balkan States, the Middle East, the Far East and the Pacific have yet to be presented.

Thus declines of frog populations are global, but it is far from clear whether an 'unknown' factor is at work in addition to a few that can be identified.

DECLINES IN AUSTRALIA

The initial assessment of declining frog populations in Australia (prepared for the Irvine workshop) was that 23 species were exhibiting declines (Tyler, 1991a). In the majority of cases the significance of the decline was unknown, i.e. whether a species was declining in numbers in only a small portion of its geographic range or whether the greater part of the range was affected.

Currently at least 34 species should be declared 'endangered', 'vulnerable', or 'insufficiently known' in conservation terms, the last category involving species reported to be declining in just a portion of their geographic range (Table 22). In addition there is a further 30 species about which the knowledge of distribution or biology is so inadequate that their conservation status cannot be determined. The existence of the latter group reflects the inadequacy of funding available for the most basic component of biological research; finding out the nature and diversity of the Australian fauna and flora.

To be positive, the small Australian herpetological community has responded to the challenge. Richards, McDonald and Alford (1993) examined the decline in populations in the tropical rainforests of northern Queensland. Their research was supported by the Wet Tropics Management Agency, which has the responsibility to care for the area.

Richards *et al.* reported that they were unable to locate *Litoria nyakalensis* and *Taudactylus rheophilus* at any site during their 1991–92 surveys and that these species have not been seen since April 1990 and January 1990 respectively. They draw attention to the fact that both are confined to upland rainforest. Hence it may be highly significant that whereas they located *Litoria nannotis*, *L. rheocola* and *Nyctimystes dayi* at lowland sites, they had disappeared from most upland sites located to the south of the Daintree River. *Taudactylus acutirostris* is a further species which is a source of concern. Because it is active by day, it is conspicuous and has been reasonably easy to locate. However it is evident that it has disappeared from a number of sites where it had occurred formerly.

In terms of explaining the declines or disappearances, Richards *et al.* (1993) drew a blank. Water samples taken at the sites failed to indicate any rise in heavy metals (which often are toxic to frogs), or other factors influencing water quality where declines were noted. Pesticide residues and temporary drought were also ruled out as possible causes.

Studies in other parts of Australia have not been accompanied by such rigorous investigations of causes. However the declines or disappearances are real and, with one exception, remain unexplained. The exception seems to have involved an undescribed species. Mr C.E. Rix (personal communication) told of a small green tree frog that lived in the reed beds that once occupied an extensive area west of the city of Adelaide. The reed beds, and the entire aquatic fauna and flora, were destroyed progressively as the land was 'reclaimed' (what an inappropriate word) for metropolitan housing. Nothing is now left.

The frog involved is unlike any species now found in South Australia. Sadly the frog known to Mr Rix became extinct before it was known to science.

But the realisation that frog populations might be vulnerable was brought home dramatically when the fabulous gastric brooding frog *Rheobatrachus silus* disappeared. The demise of this species became a focus simply because it was the subject of extensive research on gastric ulcers and accordingly, had its natural populations monitored. From a static state of abundance it simply disappeared (Tyler and Davies, 1985). Curiously the diurnal frog *Taudactylus diurnus* once abundant in the same area, disappeared at the same time (Czechura and Ingram, 1990; Czechura, 1991).

As a component of a national audit of the state of all vertebrate species the Australian National Conservation Agency (formerly the Australian National Parks and Wildlife Service), initiated an action plan to identify and highlight species requiring special conservation attention (Tyler, 1994). There is now a national awareness that steps are needed to ensure the survival of Australian frogs.

CAUSES AND POSSIBLE ACTIONS

Before anyone refers this issue to the 'too hard basket' it needs to be stated that one or two significant steps can be taken in areas where the cause of decline is obvious. For example, halting the luxury market component of frog consumption. Ecological disasters were created in India and Bangladesh as a result of decimation of frog populations to provide an export market catering for restaurants. The impact in the current major exporting nations such as Indonesia has yet to be realised.

I can find no reason to complain about restaurants providing frogs' legs that have been reared for the restaurant trade. To argue against it would be akin to equating declines of native bird populations, with the commercial activities of organisations such as Colonel Sanders' Kentucky Fried Chicken. There is also a great deal of difference between the responsibility of an affluent European who has a choice in what he or she can eat, and the needs of a New Guinean tribesman who may not. Thus I'm not suggesting that all humans halt eating frogs' legs, but rather that those who do not need to should stop. Ultimately frog farming, particularly in areas of high unemployment and low incomes, could provide a valuable economic stimulus.

Embargoes, whether voluntary or mandatory, represent just one significant step that can be taken as a conservation measure. It will alleviate any feeling of total helplessness to counter the significant frog population declines that cannot at present be explained, let alone halted or reversed.

One of the factors considered highly significant in declines of European species is the impact of acid rain. The burning of fossil fuels has produced atmospheric pollution on a very large scale, of which sulphur is a major element. It is largely that atmospheric sulphur which is responsible for altering rain to a form in which it is acidic and potentially destructive.

John Harte of the University of California has provided data that permit the interpretation that acid rain may have become an environmental whipping-boy blamed for almost every event in which it could be even remotely implicated. This conclusion does not refute the European experience, where frog breeding ponds have become so highly acidic that little life in any form can survive there. But it is clear that resorting to acid rain as the first explanation for frog declines is unjustified. There are in fact exceptional species that are confined to waters that are acidic due to natural organic sources—for example, derived from humus—and appear to have evolved because of their capacity to adapt to those conditions. An example of this curious phenomenon is in Australia where *Litoria cooloolensis* is restricted to a unique area in south-east Queensland where the water in the swamps and slow-flowing creeks looks like cold tea.

Corn, Stolzenburg and Bury (1989) concluded that acidification of water was not a major factor in declines of species in the States of Colorado and Wyoming.

Again we are forced to make a value judgment about the quality of life that we believe we have a right to expect, the impact of those expectations upon the environment, and particularly the organisms with which we try to coexist. Given the supposed aura of 'environmental awareness' it is often assumed that new chemical compounds harnessed for human benefit will be environmentally neutral. But despite the apparent rigour of the scrutiny of toxic and other properties of new compounds it is perfectly obvious that their impact can only be determined for an infinitesimal proportion of the world's fauna. Target species are those readily obtained, easily maintained in captivity and which reproduce readily there. Two or three species of frogs are common subjects, and the convenient assumption is that the remaining species will behave similarly.

Of course this attitude is utter nonsense

and, even if it were not so, the temperature at which tests are conducted by the main analytical laboratories in the northern hemisphere (usually 20–25°C) has little applicability to exposures in desert pools in southern countries where it can often exceed 40°C. In general the amount of the compound in solution is related to the temperature!

Any value judgement would conclude that the loss of any number of frogs is a small price to pay for the eradication of insects that have been vectors of human diseases, or have ravaged crops, causing death and indescribable misery to millions of people. However the point has been reached where indiscriminate applications are beginning to have a lasting impact and, in some areas, human welfare is at stake. It is in this respect that the innocent frog is being appreciated as a sensitive indicator organism of subtle environmental change. It constitutes an early, environmental warning system of benefit to humanity.

The sensitivity of frogs is largely a consequence of their dependence upon water. When eggs are laid in an aquatic environment they are exposed to any contaminant that is present. Damage is done during the early stages of egg cleavage causing either the death of the early embryo, or else the development of abnormalities that are most obvious when they affect the hands or feet (see Chapter 13).

A hotter world is a hot topic. The Greenhouse Effect is unquestionably a reality. All that we don't know is how many panes of glass are in the greenhouse: how hot will it get? Climatic data accumulated over more than a century demonstrate that, despite oscillations, there has been a definite and progressive increase—the 1980s was the hottest decade on record.

Global warming is attributed to the 10 per cent increase in atmospheric carbon dioxide that has taken place since the 1960s. Could it have in any way contributed to the decline of frogs? It is likely that the cold-adapted montane frogs could be susceptible to elevated temperatures than frogs at sea-level. At the other extreme, tadpoles of species in the tropics of northern Australia, currently exposed to temperatures in excess of 40°C (Table 8),although obviously hot-adapted, no doubt have an upper limit of survival that they may be approaching.

TABLE 22: Australian species of frogs at risk

GENUS AND SPECIES	DISTRIBUTION	CONSERVA-TION STATUS	NATURE OF CONCERN
HYLIDAE			
Litoria aurea	SE Australia	Vulnerable	localised population disappearance
Litoria brevipalmata	Eastern NSW, SE Qld.	Vulnerable	localised population disappearance
Litoria dentata	Eastern NSW, SE Qld.	Vulnerable	localised population disappearance
Litoria flavipunctata	SE Australia	Endangered	total disappearance in ACT
Litoria lesueuri	SE Australia	Vulnerable	localised population disappearance
Litoria nannotis	Qld.	Endangered	major population declines
Litoria nyakalensis	Qld.	Endangered	major population declines
Litoria pearsoniana	Eastern NSW, SE Qld.	Vulnerable	localised population disappearance
Litoria piperata	Eastern NSW	Vulnerable	localised population disappearance
Litoria rainifomis	SE Australia	Vulnerable	localised population disappearance
Litoria rheocola	Qld.	Endangered	major population decline
Litoria spenceri	SE Australia	Endangered	major population decline
Litoria verreauxi	SE Australia	Vulnerable	localised population disappearance
Nyctimystes dayi	Qld.	Endangered	major population decline
LEPTODACTYLIDAE			
Adelotu brevis	Eastern NSW, SE Qld.	Vulnerable	localised population decline
Geocrinia alba	WA	Endangered	habitat loss
Geocrinia vitellina	WA	Endangered	habitat loss
Heleioporus albopunctatus	WA	Vulnerable	localised population decline
Lechriodus fletcheri	Eastern NSW, SE Qld.	Vulnerable	localised population decline
Mixophyes balbus	Eastern NSW, SE Qld.	Vulnerable	localised population decline
Mixophyes fleayi	Eastern NSW, SE Qld.	Vulnerable	localised population decline
Mixophyes iteratus	Eastern NSW, SE Qld.	Vulnerable	localised population decline
Philoria frosti	Vic.	Vulnerable	habitat destruction
Pseudophryne australis	NSW	Vulnerable	habitat destruction
Pseudophryne bibroni	SE Australia	Vulnerable	localised population decline
Rheobatrachus silus	SE Qld.	Possibly extinct	not seen since 1981
Rheobatrachus vitellinus	Central coastal Qld.	Possibly extinct	not seen since 1985
Taudactylus acutirostris	Eastern coastal Qld.	Endangered	fatal disease feared
Taudactylus diurnus	SE Qld.	Possibly extinct	not seen since 1981
Taudactylus eungellensis	Central coastal Qld.	Endangered	few individuals remain
Taudactylus pleione	SE Qld.	Vulnerable	few individuals remain
Taudactylus rheophilus	SE Qld.	Vulnerable	few individuals remain
Uperoleia marmorata	WA	Indeterminate	not seen for 150 years
Uperoleia orientalis	NT	Indeterminate	seen once in 50 years

The definitions of conservation status indicate the level of concern. Thus, although *Uperoleia marmorata* has not been seen for 150 years, it is of 'indeterminate' status only because no-one has visited the remote site where it was first collected. It may well be quite secure. In contrast then is concern about *Rheobatrachus silus* which has not been seen for thirteen years because extensive and intensive searches have failed to relocate any individuals.

REFERENCES

Abbasi, S.A. and Soni, R. (1984). `Teratogenic effects of chromium (VI) in environment as evidenced by the impact on larvae of amphibian *Rana tigrina*: implications in the environmental management of chromium'. *Intern. J. Environ. Stud.* 23: 131-137.

Adams, N.G. (1967). '*Bufo marinus* eaten by *Rattus rattus*'. *N. Queensl. Nat.* 34: 5.

Alcala, A.C. (1957). 'Philippine notes on the ecology of the giant marine toad'. *Silliman J.* 4: 90–96.

Alexander, T.R. (1965). 'Observations on the feeding behavior of *Bufo marinus* (Linne)'. *Herpetologica*, 20(4): 255–259.

Allen, E.R. and Neill, W.T. (1956) 'Effects of marine toad toxins on man'. *Herpetologica*, 12(2): 150–151.

Anderson, R.J. and Prahlad, K.V. (1976). `The deleterious effects of fungicides and herbicides on *Xenopus laevis* embryos'. *Archiv. Environ. Contam. Toxicol.* 4: 312–323.

Anstis, M. (1974). `An introduction to the study of Australian tadpoles'. *Herpetofauna*, 7: 9–14.

Anstis, M. (1976). `Breeding biology and larval development of *Litoria verreauxi* (Anura: Hylidae)'. *Trans. R. Soc. S. Aust.* 100: 193–202.

Anstis, M. (1981). `Breeding biology and range extension for the New South Wales frog *Kyarranus sphagnicolus* (Anura: Leptodactylidae)'. *Aust. J. Herpetol.* 1: 1–9.

Australian Society of Herpetologists (1987). `Case 2531. Three works by Richard W. Wells and C. Ross Wellington: proposed suppression for nomenclatural purposes'. *Bull. zool. Nomencl.* 44(2): 116–121.

Ayre, D.J., Coster, P., Bailey, W.J. and Roberts, J.D. (1984). `Calling tactics in *Crinia georgiana* (Anura: Myobatrachidae): alternation and variation in call duration'. *Aust. J. Zool.* 32: 463–470.

Backstrom, S. (1953). `Morphogenetic effects of lithium on the embryonic development of *Xenopus*'. *Arkiv. Zool.* 6(27): 527–536.

Bailey, W.J. and Roberts, J.D. (1981). `The bioacoustics of the burrowing frog *Heleioporus* (Leptodactylidae)'. *J. Nat. Hist.* 15: 693–702.

Baldwin, P.H., Schwartz, C.H. and Schwartz, E.R. (1952). `Life history and economic status of the mongoose in Hawaii'. *J. Mammal.* 33(3): 335–356.

Bancroft, R. and Prahlad, K.V. (1973). `Effect of ethylenebis [dithiocarbamic acid] disodium salt (nabam) and ethylenebis [dithiocarbamato] maganese (maneb) on *Xenopus laevis* development'. *Teratology*, 7: 143–150.

Bando, R. (1976). 'Heavy metals concentrations (chromium, copper, manganese and lead) in tadpoles and adults of *Rana esculenta* (L)'. *Inst. Ital. Idrobiol. Dolt. Marco Marchi.* 33: 325–344.

Banks, C.B., Birkett, J.R., Dunn, R.W. and Martin, A.A. (1983). 'Development of *Litoria infrafrenata* (Anura: Hylidae)'. *Trans. R. Soc. S. Aust.* 107: 197–200.

Baringa, M. (1990). 'Where have all the froggies gone?' *Science*, 247, 1033–1034.

Barker, J. and Grigg, G. (1977). *A field guide to Australian frogs*. Rigby, Adelaide.

Barker, J., Grigg, G. and Tyler, M.J. (1994). *A field guide to Australian frogs*. Surrey Beatty & Sons, Chipping Norten, NSW.

Beattie, R.C. (1980). 'A physico–chemical investigation of the jelly capsules surrounding eggs of the common frog *(Rana temporaria temporaria)*'. *J. Zool., Lond.* 190: 1–25.

Beatty, P.W., Holscher, M.A. and Neal, R.A. (1976). 'Toxicity of 2, 3, 7, 8–tetrachlorodibenzo–p–dioxin in larval and adult forms of *Rana catesbeiana*'. *Bull. Environ. Contam. Toxicol.* 16: 578–581.

Bell, A.F. (1940). 'The giant toad in Queensland Australia'. *Agric. J.* (Fiji), 11(2): 55.

Best, S.M. (1973). 'Some organochlorine pesticide residues in wildlife of the Northern Territory, Australia, 1970–71'. *Aust. J. Biol. Sci.* 26: 1161–1170.

Bettinger, H.F. and O'Loughlin, I. (1950). 'The use of the male toad, *Bufo marinus,* for pregnancy tests'. *Med. J. Aust.*, July 8, 1950: 40–42.

Bimber, D.L. and Mitchell, R.A. (1978). 'Effects of Diquat on amphibian embryo development'. *Ohio J. Sci.* 78: 50–51.

Birge, W.J., Black, J.A., Westerman, A.G., Francis,

P.C. and Hudson, J.E. (1977). 'Embryopathic effects of waterborne and sediment–accumulated cadmium, mercury and zinc on reproduction and survival of fish and amphibian populations in Kentucky'. *Univ. Kentucky Water Resour. Res. Inst. Res. Rep.* (100).

Birge, W.J. and Just, J.J. (1973). 'Sensitivity of vertebrate embryos to heavy metals as a criterion of water quality'. *Univ. Kentucky Water Resour. Inst. Res. Rep.* (61).

Birks, P.R. and Olsen, A.M. (1987). 'Pesticide concentrations in some South Australian birds and other fauna'. *Trans. R. Soc. S. Aust.* 111(2): 67–77.

Bishop, C.A. and Pettit, K.E. (1993). 'Declines in Canadian amphibian populations: designing a national monitoring strategy'. Proceedings of a workshop sponsored by the Canadian Wildlife Service (Ontario Region) and the Metropolitan Toronto Zoo, held in Burlington, Ontario, 5–6 October, 1991. *Occasional paper Canadian Wildlife Service*, (76), 1–120.

Blanchard, F.N. (1929). 'Re–discovery of *Crinia tasmaniensis*'. *Aust. Zool.* 5: 324–325.

Blinov, V.A. (1962). 'Sensitivity of amphibian embryos to X–rays at various stages of developrnent'. *Prob. Radiobiol.* 1: 133–153.

Boice, R. and Boice, C. (1971). 'Interspecific competition in captive *Bufo americanus* toads'. *J. Biol. Psych.* 12: 32–36.

Bradford, D.F. and Seymour, R.S. (1985). 'Energy conservation during the delayed–hatching period in the frog *Pseudophryne bibroni*'. *Physiol. Zool.* 58: 491–496.

Brattstrom, B.H. (1970). 'Thermal acclimation in Australian amphibians'. *Comp. Biochem. Physiol.* 35: 69–103.

Breder, C.M. (1946). 'Amphibians and reptiles of the Rio Chacunaque Drainage, Darien, Panama, with notes on their life–histories and habits'. *Bull. Amer. Mus. Nat. Hist.* 86: 375– .

Brooks, A.J. (1983). *Atlas of Australian Anura*. Dept of Zoology, University of Melbourne, Parkville. Publ. No. 7.

Brooks, J.A. (1981). 'Otolith abnormalities in *Limnodynastes tasmaniensis* tadpoles after embryonic exposure to the pesticide dieldrin'. *Environ. Pollut.* 25(1): 19–25.

Browne, C.L. and Dumont, J.N. (1979). 'Toxicity of selenium to developing *Xenopus laevis* embryos'. *J. Toxicol. Environ. Health,* 5: 699–709.

Bull, C.M. (1978). 'The position and stability of a hybrid zone between the Western Australian frogs *Ranidella insignifera* and *R. pseudinsignifera*'. *Aust. J. Zool.* 26: 305–322.

Bustuoabad, O.D., Herkovits, J. and Pisano, A. (1977). 'Different sensitivity to lithium ion during the segmentation of *Bufo arenarum* eggs'. *Acta Embryol. Exper.* (3): 271–282.

Buzacott, J.H. (1939). Toads and fowls. Cane Pests Board Conf. Proc. 28.

Byrne, A.R., Kosta, L. and Stegnar, P. (1975). 'The occurrence of mercury in Amphibia'. *Environmental Letters* 8(2): 147–155.

Calaby, J.H. (1956), 'The food habits of the frog, *Myobatrachus gouldii* (Gray)'. *West. Aust. Nat.* 5(4): 93–96.

Calaby, J.H. (1960). 'A note on the food of Australian desert frogs'. *West. Aust. Nat.* 7(3): 79–80.

Cappo, M.C. (1986). *Frogs as predators of organisms of aquatic origin in the Magela Creek System, Northern Territory*. (MSc Thesis). Dept of Zoology, University of Adelaide. (Unpublished).

Carter, D.B. (1979). 'Structure and function of the subcutaneous lymph sacs in the Anura (Amphibia)'. *J. Herpetol.* 13(3): 321–327.

Cassels, M. (1970). 'Another predator of the Cane Toad'. *N. Qld. Nat.* 37(151): 6.

Cogger, H.G. (1975). *Reptiles and amphibians of Australia*. Reed, Sydney.

Cogger, H.G., Cameron, E.E. and Cogger, H.M. (1983). *Zoological catalogue of Australia, I. Amphibia and Reptilia*. 313 pp. Aust. Govt Publ. Service, Canberra.

Connell, D.W. (1974). *Water pollution: causes and effects in Australia*. University of Queensland Press, St Lucia.

Cooke, A.S. (1970). 'The effect of pp – DDT on tadpoles of the common frog *(Rana temporaria)*'. *Environ. Pollut.* 1: 57

Cooke, A.S. (1972). 'The effects of DDT, Dieldrin and 2, 4–D on amphibian spawn and tadpoles'. *Environ. Pollut.* 3: 51–68.

Cooke, A.S. (1973a). 'Response of *Rana temporaria* tadpoles to chronic doses of pp – DDT'. *Copeia* 1973: 647–652.

Cooke, A.S. (1973b). 'The effects of DDT, when used as a mosquito larvicide, on tadpoles of the frog *Rana temporaria*. *Environ. Pollut.* 5: 258–273.

Cooke, A.S. (1974). 'The effects of pp – DDT on adult frogs *(Rana temporaria)*'. *Brit. J. Herpet.* 5:

390–396.

Cooke, A.S. (1977). 'Effects of field applications of the herbicides diquat and dichlobenil on amphibians'. *Environ. Pollut.* 12: 43–50.

Cooke, A.S. (1979). 'The influence of rearing density on the subsequent response to DDT dosing for tadpoles of the frog *Rana temporaria*'. *Bull. Environ. Contam. Toxicol.* 21: 837–841.

Cooke, A.S. (1981). 'Tadpoles as indicators of harmful levels of pollution in the field'. *Environ. Pollut.* (Ser. A) 25: 123–133.

Corben, C.J. and Ingram, G.J. (1987). 'A new barred river frog (Myobatrachidae, *Mixophyes*)'. *Mem. Qld Mus.* 25: 233–237.

Corn, P.S., Stolzenburg, W. and Bury, R.B. (1989). 'Acid precipitation studies in Colorado and Wyoming: interim report of surveys of montane anphibians and water chemistry'. *U.S. Fish. Wildl. Serv. Biol. Rep.* 80 (40.26), 1–56.

Costantini, F. and Andreoli, V. (1972). 'Erbicidi e fauna. Nota 1: Effetti degli erbicidi sullo svilluppo embrionale e post embrionale di *Bufo vulgaris* L'. *Ann. Fac. Agrar. Univ. Perugia*, 27: 227–236.

Costantini, F. and Panella, F. (1975). 'Erbicidi e fauna. Nota III. Effetti degli erbicidi sullo suilluppo embrionale e postembrionale di *Bufo vulgaris* L. e modificazioni enzimatiche nelle larve'. *Ann. Fac. Agrar. Univ. Perugia*, 29: 421–440.

Covacevich, J. (1974). 'An unusual aggregation of snakes following major flooding in the Ipswich–Brisbane area, south-eastern Queensland'. *Herpetofauna*, 7(1): 21–24.

Covacevich, J. and Archer, M. (1975). 'The distribution of the Cane Toad, *Bufo marinus*, in Australia and its effects on indigenous vertebrates'. *Mem. Qld. Mus.* 17(2): 305–310.

Crook, G.A. and Tyler, M.J. (1981). 'Structure and function of the tibial gland of the Australian frog *Limnodynastes dumerilii* Peters'. *Trans. R. Soc. S. Aust.* 105: 49–52.

Crossland, M.R. and Richards, S.J. (1993). 'The tadpole of the Australopapuan frog *Litoria nigrofrenata* (Gunther, 1867) (Anura: Hylidae)'. *Trans. R. Soc. S. Aust.* 117: 109–110.

Czechura, G.V. (1991). 'The Blackall Ranges: frogs, reptiles and fauna conservation' pp. 311–324 *In* 'The Rainforest Legacy'. Australian Rainforests Study. Vol. 2, Special Heritage Publication Series. Aust. Gov. Publ. Service, Canberra.

Czechura, G.V. and Ingram, G.J. (1990). '*Taudactylus diurnus* and the case of the disappearing frogs'. *Mem. Qld. Mus.* 29, 361–365.

Czechura, G.V., Ingram, G.J. and Liem, L.S. (1987). 'The genus *Nyctimystes* (Anura: Hylidae) in Australia'. *Rec. Aust. Mus.* 39(5): 333–338.

Daugherty, D.H. and Maxson, L.R.. (1982). 'A biochemical assessment of the evolution of myobatrachille frogs'. *Herpetologica,* 38: 341–348.

Davies, M. (1983). 'Skeleton' *In* M.J. Tyler (Ed.) *The Gastric Brooding Frog.* Croom Helm, London & Canberra.

Davies, M. (1984). 'Osteology of the myobatrachine frog *Arenophryne rotunda* Tyler (Anura: Leptodactylidae) and comparisons with other myobatrachine genera'. *Aust. J. Zool.* 32: 789–802.

Davies, M. (1989). 'Developmental biology of the Australopapuan hylid frog *Litoria eucnemis* (Anura: Hylidae)'. *Trans. R. Soc. S. Aust.* 113(4): 215–220.

Davies, M. (1991). 'Description of the tadpoles of some Australian limnodynastine leptodactylid frogs'. *Trans. R. Soc. S. Aust.* 115(2): 67–76.

Davies, M. and Littlejohn, M.J. (1986). 'Frogs of the genus *Uperoleia* Gray (Anura: Leptodactylidae) in south-eastern Australia'. *Trans. R. Soc. S. Aust.* 109(3): 111–143.

Davies M., Mahony, M. and Roberts, J.D. (1985). 'A new species of *Uperoleia* (Anura: Leptodactylidae) from the Pilbara region, Western Australia'. *Trans. R. Soc. S. Aust.* 109(3): 103–108.

Davies, M., Martin, A.A. and Watson, G.F. (1983). 'Redefinition of the *Litoria latopalmata* species group (Anura: Hylidae)'. *Trans. R. Soc. S. Aust.* 107(2): 87–108.

Davies, M. and McDonald, K.R. (1979). 'A study of intraspecific variation in the green tree frog *Litoria chloris* (Boulenger) (Hylidae)'. *Aust. Zool.* 20: 347–359.

Davies, M. and McDonald, K.R. (1985). 'A redefinition of *Uperoleia rugosa* (Andersson) (Anura: Leptodactylidae)'. *Trans. R. Soc. S. Aust.* 109(2): 37–42.

Davies, M., McDonald, K.R. and Adams, M. (1986). 'A new species of green tree frog (Anura: Hylidae) from Queensland, Australia'. *Proc. R. Soc. Vict.* 98(2): 63–71.

Davies, M., McDonald, K.R. and Corben, C. (1986). 'The genus *Uperoleia* Gray (Anura: Leptodactylidae) in Queensland. Australia'. *Proc. R. Soc. Vict.* 98(4): 147–188.

Davies, M. and Richards, S.J. (1990). 'Developmental biology of the Australian hylid frog *Nyctimystes dayi* (Gunther)'. *Trans. R. Soc. S. Aust.* 114: 207–211.

Davies, M., Watson, G.F. and McDonald, K.R. (1992). 'Redefinition of *Uperoleia littlejohni* Davies, McDonald & Corben (Anura: Leptodactylidae: Myobatrachinae).' *Trans. R. Soc. S. Aust.* 116(4): 137–139.

Davies, M., Watson, G.F., Mcdonald, K.R., Trenery, M.P. and Werren, G. (1993).'A new species of *Uperoleia* (Anura: Leptodactylidae; Myobatrachinae) from northeastern Australia'. *Mon. Qld. Mus.* 33(1): 167–174.

de la Lande, I.S., O'Brien, P.E., Shearman, D.J.C., Taylor, P. and Tyler, M.J. (1984). 'On the possible role of prostaglandin E$_2$ in intestinal stasis in the gastric brooding frog *Rheobatrachus silus*'. *Aust. J. Exp. Biol. Med. Sci.* 62: 317–323.

Delrio, G., Citarella, F. and d'Istria, M. (1980). 'Androgen receptor in the thumb pad of *Rana esculenta*: dynamic aspects'. J. *Endocr.* 84: 279–282.

Deschiens, R., Floch, H. and Le Corroller, Y. (1965). 'Les molluscicides cuivreux dans la prophylaxie des bilharzioses'. *Bull. Org. mond. Sante,* 33: 73–88.

Dexter, R.R. (1932). 'The food habits of the imported toad *Bufo marinus* in the sugar cane sections of Puerto Rico'. *Bull. Int. Soc. Sugar Cane Technol.* (74): 1–6.

Dial, N.A. (1976). 'Methylmercury: teratogenic and lethal effects in frog embryos'. *Teratology,* 13: 327–334.

Dial, N.A. and Bauer, C.A. (1984). 'Teratogenic and lethal effects of Paraquat on developing frog embryos *(Rana pipiens)*'. *Bull. Environ. Contam. Toxicol.* 33: 592–597.

Dial, N.A. and Bauer Dial, C.A. (1987). 'Lethal effects of Diquat and Paraquat on developing frog embryos and 15–day–old tadpoles, *Rana pipiens*'. *Bull. Environ. Contam. Toxicol.* 38: 1006–1011.

Dilling, W.J. and Healey, C.W. (1925). 'Influence of lead and the metallic ions of copper, zinc, thorium, beryllium and thallium on the germination of frogs' spawn and on the growth of tadpoles'. *Ann. Appl. Biol.* 13: 177–188.

Domm, A.J. and Janssens, P.A. (1971). 'Nitrogen metabolism during development in the corroboree frog, *Pseudophryne corroboree* Moore'. *Comp. Biochem. Physiol.* 38A: 163–173.

Donnellan, S.A., Mahony, M.J. and Davies, M. 'A new species of *Mixophyes* (Anura: Leptodactylidae) and first record of the genus in New Guinea'. *Copeia.*

Dubois, A. (1979). 'Anomalies and mutation in natural populations of the *Rana "esculenta"* complex (Amphibia, Anura)'. *Mitt. Zool. Mus. Berlin,* 55: 59–87.

Dubois, A. (1983). 'Classification et nomenclature supragenerique des amphibiens anoures'. *Bull. Mens. Soc. Linn. Lyon* 52(9): 270–276.

Dubois, A. (1984). 'La nomenclature supragenerique des amphibiens anoures'. *Mem. Mus. Nat. Hist. Nat. Paris. Ser. A. Zool.* 131: 1–64.

Dubois, A. (1985). 'Miscellanea nomenclatorica batrachologica (VII)'. *Alytes* 4(2): 61–78.

Duellman, W.E. (1975). 'On the classification of frogs'. *Occ. Pap. Mus. Nat. Hist. Univ. Kansas* (42): 1–14.

Duellman, W.E. and Trueb, L. (1986). *Biology of Amphibians.* McGraw–Hill, New York.

Dutta, S.K. and Mohanty–Hejmadi, P. (1978). 'Life history and pesticide susceptible embryonic stages of the Indian bull frog *Rana tigrina* Daudin'. *Indian. J. Exp. Biol.* 16: 727–729.

Easteal, S. (1981). 'The history of introductions of *Bufo marinus* (Amphibia: Anura); a natural experiment in evolution'. *Biol. J. Linn. Soc.* 16: 93–113.

Easteal, S., Van Beurden, E. K., Floyd, R. B. and Sabath, M.D. (1985). 'Continuing geographical spread of *Bufo marinus* in Australia: range expansion between 1974 and 1980'. *J. Herpetol.* 19(2): 185–188.

Edery, H. and Schatzberg–Porath, G. (1960). 'Studies on the effect of organophosphorus insecticides on amphibians'. *Arch. int. pharmacodyn.* 124(102): 212–224.

Ehmann, H. and Swan, G. (1985). 'Reproduction and development in the marsupial frog, *Assa darlingtoni* (Leptodactylidae, Anura)'. *In* G. Grigg, R. Shine and H. Ehmann (Eds) *Biology of Australasian frogs and reptiles.* Surrey Beatty & Sons, Chipping Norton, NSW.

Ely, C.A. (1944). 'Development of *Bufo marinus* larvae in dilute sea water'. *Copeia* 1956(4): 256.

Emerson, S.B. and Diehl, D. (1980). 'Toe pad morphology and mechanisms of sticking in frogs'. *Biol. J. Linn. Soc.* 13(3): 199–216.

Estes, R. (1984). 'Fish, amphibians and reptiles from the Etadunna Formation, Miocene of South Australia'. *Aust. Zool.* 21: 335–343.

Farris, J.S. Kluge, A.G. and Mickevich, M.F. (1982). 'Phylogenetic analysis, the monothetic group method, and myobatrachid frogs'. *Syst. Zool.* 31(3): 317–327.

Ferguson, D.E. and Gilbert, C.C. (1967). 'Tolerance of three species of anuran amphibians to five chlorinated hydrocarbon insecticides'. *Miss. Afad. Sci. J.* 13: 135–138.

Fingal, W. and Kaplan, H.M. (1963). 'Susceptibility of *Xenopus laevis* to copper sulphate'. *Copeia* 1963(1): 155–156.

Fischer, J.-L. (1973). 'Polydactylie faible chez la grenouille rousse'. *Bull. Mensuel Soc. Linn. Lyon,* 42(1): 1–4.

Fischer, J.-L. (1977). 'Un mime morphologique de la polydactylie faible: la fissuration de pha-langes distales chez *Rana temporaria* (Amphibiens, Anoures)'. *Bull. Soc. Linn. Lyon,* 46(5): 143–146.

Fleet, R.R., Clark, D.R. and Plapp, F.W. (1972). 'Residues of DDT and dieldrin in snakes from two Texas agro–systems'. *Bioscience,* 22: 664–665.

Forbes, T.W., Abbott, P.S. and Hamre, C.J. (1949). 'An action potential study of taste responses in the toad'. (Abstract). *Hawaii. J. Sci.* 1949: 9–10.

Freeland, W.J. (1985). 'The need to control cane toads'. *Search* 16(7–8): 211–215.

Freeland, W.J. and Martin, K.C. (1985). 'The rate of expansion by *Bufo marinus* in northern Australia'. *Aust. Wildl. Res.* 12: 555–559.

Frith, H.J. (1979). *Wildlife Conservation.* Angus & Robertson. Sydney.

Froggatt, W.W. (1936). 'The introduction of the great Mexican toad *Bufo marinus* into Australia'. *Aust. Nat.* 9: 163–164.

Frost, D.R. (1985). (Ed.) *Amphibian species of the world.* Allen Press and Association of Systematics Collections, Lawrence, Kansas.

Gabor, M., Puscariu, F. and Deac, C. (1973). 'Action teratogene des aflatoxines sur les tetards'. *Arch. Roum. Path. exp. Microbiol.* 32: 269–275.

Galli–Mainini, C. (1947). 'Pregnancy test using the male Batrachia'. *J. Amer. Med. Assoc.* 138: 121.

Gans, C. (1985). 'Comment on two checklists'.

Herpetol. Rev. 16(1): 6–7.

Gartside, D.F., Littlejohn, M.J. and Watson, G.F. (1979). 'Structure and dynamics of a narrow hybrid zone between *Geocrinia laevis* and *G. victoriana* (Anura: Leptodactylidae) in south–eastern Australia'. *Heredity,* 43(2): 165–177.

Ghate, H.V. and Mulherkar, L. (1980). 'Effect of mercuric chloride on embryonic development of the frog *Microhlyla ornata*'. *Indian J. Exp. Biol.* 18: 1094–1096.

Gollman, B. (1991). 'A developmental table of *Crinia signifera* Girard, 1853 (Anura, Myobatrachinae)'. *Alytes,* 9(2), 51–58.

Gollman, B. and Gollman, G. (1991). 'Embryonic development of the myobatrachine frogs *Geocrinia laevis, Geocrinia victoriana,* and their natural hybrids'. *Amphibia–Reptilia* 12, 103–110.

Goodacre, W.A. (1947). 'The giant toad *(Bufo mari-nus)* an enemy of bees'. *Agric. Gaz. NSW,* 58: 374–375.

Gorham, S.W. (1968). 'Fiji frogs. Life history data from field work'. *Zoolog. Bietr.* 14: 427–446.

Gradwell, N. (1975). 'The clinging mechanism of *Pseudophryne bibroni* (Anura: Leptodactylidae) to an alga on glass'. *Trans. R. Soc. S. Aust.* 99: 31–34.

Green, D.M. (1979). 'Treefrog toe pads: compara-tive surface morphology using scanning elec-tron microscopy'. *Can. J. Zool.* 57(10): 2033–2046.

Green, D.M. (1981). 'Adhesion and toe–pads of treefrogs'. *Copeia,* 1981(4): 790–796.

Grosch, D.S. and Hopwood, L.E. (1979). *Biological effects of radiations.* 2nd Edition. Academic Press, New York.

Gruneberg, H. (1963) . *The pathology of develop-ment.* Blackwell, London.

Gruneberg, J. (1956). 'Genetical studies on the skeleton of the mouse. XVIII. Three genes for syndactylism'. (With an appendix by D.S. Falconer). *J: Genet.* 54: 113–145.

Gunn, R.H., Galloway, R.W., Walker, J., Nix, H.A., McAlpine, J.R. and Richardson, D.P. (1972). 'Shoalwater Bay area, Queensland'. *Tech. Mem.* 72/10. CSIRO Div. Land Res., Canberra.

Hall, R.J. and Kolbe, E. (1980). 'Bioconcentration of organophosphorus pesticides to hazardous levels'. *J. Toxicol. Environ. Health* 6: 853– 860.

Hall, R.J. and Swineford, D. (1979). 'Uptake of methoxychlor from food and water by the

American toad *(Bufo americanus)'. Bull. Environ. Contam. Toxicol.* 23: 335–337.

Hall, R.J. and Swineford, D. (1980). 'Toxic effects of endrin and toxaphene on the southern leopard frog *Rana sphenocephala'. Environ. Pollut.* (Ser. A), 23: 53–65.

Hamilton, L. (1969). 'Changes in survival after X–irradiation of *Xenopus* embryos at different phases of the cell cycle'. *Radiat. Res.* 37: 173–180.

Harding, K.A. (1982). 'Courtship display in a Bornean frog'. *Proc. biol. Soc. Wash.* 95(3): 621–624.

Harfenist, A., Power, T., Clark, K.C. and Peakall, D.B. (1989). 'A review and evaluation of the amphibian toxicological literature'. *Canadian Wildl. Serv. Tech. Rep. Ser.* (61), 1–222.

Harri, M.N.E., Laitinen, J. and Valkama, E.–L. (1979). 'Toxicity and retention of DDT in adult frogs, *Rana temporaria* L'. *Environ. Pollut.* 19: 45–55.

Harrison, L. (1922). 'On the breeding habits of some Australian frogs'. *Aust. Zool.* 3(1): 17–34.

Harrison, L. H. (1927). 'Notes on some Western Australian frogs, with descriptions of new species'. *Rec. Aust. Mus.* 15: 277–287.

Harrison, P.A. and Littlejohn, M.J. (1985). 'Diphasy in the advertisement calls of *Geocrinia laevis* (Anura: Leptodactylidae): vocal responses of males during field playback experiments'. *Behav. Ecol. Sociobiol.* 18: 67–73.

Hattori, K. (1974). 'Toxicity of difenphos (Abate) in the Amphibia'. *Rep. Hokkaido Inst. Pub. Health* (24): 152–154.

Hazelwood, E. (1970). 'Frog pond contaminated'. *Brit. J. Herpet.* 4: 177–184.

Heatwole, H. (1985). [No title.] *Herpetol. Rev.* 16(1): 6.

Henle, K. (1981). 'A unique case of malformations in a natural population of the green toad *(Bufo viridis)* and its meaning for environmental politics'. *Brit. Herpet. Soc. Bull.* (4): 4849.

Heyer, W.R. and Liem, D.S. (1976). 'Analysis of the intergeneric relationships of the Australian frog family Myobatrachidae'. *Smithson. Contrib. Zool.* (233): 1–29.

Hillenius, D. (1959). 'Ein zeueiter fall der "Anomalie P" (Rostand) bei *Rana esculenta* Linne'. *Med. Klin. Munich,* 54: 3–7.

Horton, P. (1982a). 'Diversity and systematic significance of anuran tongue musculature'. *Copeia,* 1982(3): 595–602.

Horton, P. (1982b). 'Precocious reproduction in the Australian frog *Limnodynastes tasmaniensis'. Herpetologica,* 38: 486–489.

Horton, P. and Tyler, M.J. (1982). 'The female reproductive system of the Australian gastric brooding frog, *Rheobatrachus silus* (Anura: Leptodactylidae)'. *Aust. J. Zool.* 30: 857–863.

Humphries, R.B. (1979). *Dynamics of a breeding frog community.* (Ph. D. thesis). Australian National University, Canberra. (Unpublished).

Hutchings, R.W. (1979). 'A native predator of the cane toad *(Bufo marinus)'. N. Queensl. Nat.* 45(174): 4–5.

Hutchinson, M.N. and Maxson, L.R. (1987a). 'Phylogenetic relationships among Australian tree frogs (Anura: Hylidae: Pelodryadinae): an immunological approach'. *Aust. J. Zool.* 35: 61–74.

Hutchinson, M.N. and Maxson, L.R. (1987b). 'Biochemical studies on the relationships of the gastric–brooding frogs, genus *Rheobatrachus'. Amphibia–Reptilia,* 8: 1–11.

Inger, R.F. (1954). 'Systematics and zoogeography of Philippine Amphibia'. *Fieldiana: Zool,* 33(4): 181–531.

Inger, R.F. (1966). 'The systematics and zoogeography of the Amphibia of Borneo'. *Fieldiana: Zool.* 52: 1–492.

Ingram, G. (1980). 'A new frog of the genus *Taudactylus* (Myobatrachidae) from mid–eastern Queensland with notes on the other members of the genus'. *Mem. Qld. Mus.* 20: 111–119.

Ingram. G.J., Anstis, M. and Corben, C.J. (1975). 'Observations on the Australian leptodactylid frog, *Assa darlingtoni'. Herpetologica,* 31: 425–429.

Ingram, G.J. and Corben, C.J. (1975). 'A new species of *Kyarranus* (Anura: Leptodactylidae) from Queensland, Australia'. *Mem. Qld. Mus.* 17: 335–339.

Ingram, G.J., Corben, C. and Hosmer, W. (1982). '*Litoria revelata*: a new species of tree–frog from eastern Australia'. *Mem. Qld. Mus.* 20(3): 635–637.

Jaskoski, B.J. and Kinders, R.J. (1974). 'DDT and methoxychlor in a frog population in Cook County'. *Trans. Ill. State Acad. Sci.* 67: 341–344.

Johnson, C.R. (1969). 'Water absorption response of some Australian anurans'. *Herpetologica,* 25(3): 171–172.

Johnson. C.R. (1976). 'Herbicide toxicities in some Australian anurans and the effect of subacute dosages on temperature tolerance'. *Zool. J. Linn. Soc.* 59: 79–83.

Jolly, A.L., Avault, J.W., Koonce, K.L. and Graves, J.B. (1978) . 'Acute toxicity of Permethrin to several aquatic animals'. *Trans. Amer. Fish. Soc.* 107: 825–827.

Jolly, A. L., Graves, J.B., Avault, J.W. and Koonce, K.L. (1977). 'Effects of a new insecticide on aquatic animals'. *La Agric.* 21: 3–16.

Jordan, M., Rzehak, K. and Maryanska, A. (1977). 'The effect of two pesticides, Miedzian 50 and Gesagard 50 on the development of tadpoles of *Rana temporaria*'. *Bull. Environ. Contam. Toxicol.* 17: 349–354.

Judd, F.W. (1977). 'Toxicity of monosodium methanearsonate herbicide to Couch's spadefoot toad, *Scaphiopus couchi*'. *Herpetologica,* 33: 44–46.

Kaplan, H.M. and Overpeck, J.G. (1964). 'Toxicity of halogenated hydrocarbon insecticides for the frog, *Rana pipiens*'. *Herpetologica,* 20: 163–169.

Kaplan, H.M. and Yoh, L. (1961). 'Toxicity of copper for frogs'. *Herpetologica,* 17(2): 131–135.

Kerr, J.F.R., Harman. B. and Searle. J. (1974). 'An electronmicroscope study of cell deletion in the anuran tadpole tail during spontaneous metamorphoseis with special reference to apoptosis of striated muscle fibres'. *J. Cell Sci.* 14: 571–585.

King, M. and Miller, J. (1985). [No title.] *Herpetol. Rev.* 16(1): 4–5.

Krakauer, T. (1970). 'Tolerance limits of the toad, *Bufo marinus,* in South Florida'. *Comp. Biochem, Physiol.* 33: 15–26.

Lande, S.P. and Guttman, S.I. (1973). 'The effects of copper sulfate on the growth and mortality rate of *Rana pipiens* tadpoles'. *Herpetologica* 29: 22–27.

Laurent, R.F. (1979). 'Esquisse d'une phylogenese des anoures'. *Bull. Soc. Zool. France* 104(4): 397–422.

Laurent, R.F. (1985). 'Sur la classification et la nomenclature des amphibiens'. *Alytes* 4(3): 119–120.

Laurent, R.F. (1986). 'Sous–Classe des lissamphibiens (Lissamphibia) Systematique'. *In* P.P. Grasse and M. Delsol (Eds) *Traite de Zoologie,* 14 Fasc. 1–B, Batraciens. Masson, Paris.

Lee, A.K. (1967). 'Studies in Australian Amphibia. II. Taxonomy, ecology and evolution of the genus *Heleioporus* Gray (Anura: Leptodactylidae)'. *Aust. J. Zool.* 15: 367–439.

Leong, A.S.-Y., Tyler, M.J. and Shearman, D.J.C. (1986). 'Gastric brooding: a new form in a recently discovered Australian frog of the genus *Rheobatrachus*'. *Aust. J. Zool.* 34: 205–209.

Lever, R.J.A.W. (1938). 'The giant toad—distribution. diet and development'. *Agric. J.* (Fiji), 9(2): 28.

Lever, R.J.A.W. (1944). 'Food of the Giant Toad'. *Agric. J.* (Fiji), 15(1): 28.

Licht, L.E. (1976). 'Time course of uptake, elimination, and tissue levels of [14C] DDT in wood–frog tadpoles'. *Can. J. Zool.* 54:355–360.

Liem, D.S. (1974a). 'A review of the *Litoria nannotis* species group, and a description of a new species of *Litoria* from Northern Queensland, Australia (Anura: Hylidae)'. *Mem. Qld. Mus.* 17(1): 151–168.

Liem, D.S. (1974b). 'A new species of the *Litoria bicolor* species group from southeast Queensland, Australia (Anura: Hylidae)'. *Mem. Qld. Mus.* 17: 169–174.

Liem, D.S. and Hosmer, W. (1973). 'Frogs of the genus *Taudactylus* with descriptions of two new species (Anura: Leptodactylidae)'. *Mem. Qld. Mus.* 16: 435–457.

Liem, D.S. and Ingram, G.J. (1977). 'Two new species of frogs (Anura: Myobatrachidae, Pelodryadidaoe) [sic] from Queensland and New South Wales'. *Vict. Nat.* 94: 255–262.

Linacre, E. and Hobbs, J. (1977). *The Australian climatic environment.* Wiley, Brisbane.

Lindgren, E. and Main, A.R. (1961). 'Natural history notes from Jigalong. IV. Frogs'. *West. Aust. Nat.* 7(8): 193–195.

Littlejohn, M.J. (1957). 'A new species of frog of the genus *Crinia*'. *West. Aust. Nat.* 6(1): 18–23.

Littlejohn, M.J. (1958). 'A new species of frog of the genus *Crinia* Tschudi from south–eastern Australia'. *Proc. Linn. Soc. NSW,* 83(2): 222–226.

Littlejohn, M.J. (1959). 'Call structure in two genera of Australian burrowing frogs'. *Copeia* 1959(3): 266–270.

Littlejohn, M.J. (1963). 'The breeding biology of the Baw Baw frog *Philoria frosti* Spencer'. *Proc. Linn. Soc. NSW,* 88: 273–276.

Littlejohn, M.J. (1964). 'Geographic isolation and mating call differentiation in *Crinia signifera*'. *Evolution,* 18: 262–266.

Littlejohn, M.J. (1965). 'Premating isolation in the *Hyla ewingi* complex (Anura: Hylidae)'. *Evolution,* 19: 234–243.

Littlejohn, M.J. (1968). 'Frog calls and the species problem'. *Aust. Zool.* 14(3): 259–264.

Littlejohn, M.J. (1969). 'Acoustic interaction between two species of leptodactylid frogs'. *Anim. Behav.* 17(4): 785–791.

Littlejohn, M.J. (1976). 'The *Litoria ewingi* complex (Anura: Hylidae) in south–eastern Australia. IV. Variation in mating-call structure across a narrow hybrid zone between *L. ewingi* and *L. paraewingi'. Aust. J. Zool.* 24: 283–293.

Littlejohn, M.J (1982). *'Litoria ewingi* in Australia: a consideration of indigenous populations, and their interactions with two closely related species'. *In* D. G. Newman (Ed.) *New Zealand Herpetology.* N.Z. Wildl. Serv. Occ. Publ. (2): 113–135.

Littlejohn, M.J. (1982). 'Vocal communications of frogs'. *Anima,* (114): 65–69.

Littlejohn, M.J. and Harrison, P.A. (1981). 'Acoustic communication in *Geocrinia virtoriana* (Anura: Leptodactylidae)'. *In* C.B. Banks and A.A. Martin (Eds) *Proc. Melbourne Herp. Symp.* Zoological Board of Victoria, Melbourne.

Littlejohn, M.J. and Harrison, P.A. (1985). 'The functional significance of the diphasic advertisement call of *Geocrinia victoriana* (Anura: Leptodactylidae)'. *Behav. Ecol. Sociobiol.* 16: 363–373.

Littlejohn, M.J., Harrison, P.A. and MacNally, R.C. (1985). 'Interspecific acoustic interactions in sympatric populations of *Ranidella signifera* and *R. parinsignifera* (Anura: Leptodactylidae)'. *In* G. Grigg, R. Shine and H. Ehmann (Eds) *The biology of Australasian frogs and reptiles.* Surrey Beatty & Sons, Chipping Norton, NSW.

Littlejohn, M.J., Loftus–Hills, J.J., Martin, A.A. and Watson, G.F. (1972). 'Amphibian fauna of Victoria. Confirmation of the records of *Litoria (–Hyla) citropa* (Tschudi) in Gippsland'. *Vict. Nat.* 89(2): 51–54.

Littlejohn, M.J. and Martin, A.A. (1964). 'The *Crinia laevis* complex (Anura: Leptodactylidae) in south–eastern Australia'. *Aust. J. Zool.* 12: 70–83.

Littlejohn, M.J. and Martin, A.A. (1965). 'A new species of *Crinia* (Anura, Leptodactylidae) from South Australia'. *Copeia* 1965: 319–324.

Littlejohn, M.J. and Martin, A.A. (1967). 'The rediscovery of *Heleioporus australiacus* (Shaw) (Anura: Leptodactylidae) in eastern Victoria'. *Proc. R. Soc. Vic.* 80(1): 31–36.

Littlejohn, M.J. and Martin, A.A. (1974). 'The Amphibia of Tasmania'. *In* W.D. Williams (Ed.) *Biogeography and Ecology in Tasmania.* W. Junk, The Hague.

Littlejohn, M.J. and Roberts, J.D. (1975). 'Acoustic analysis of an intergrade zone between two call races of the *Limnodynastes tasmaniensis* complex (Anura: Leptodactylidae) in south–eastern Australia'. *Aust. J. Zool.* 23: 113–122.

Littlejohn, M.J. and Watson, G.F. (1973). 'Mating–call variation across a narrow hybrid zone between *Crinia laevis* and *C. victoriana* (Anura: Leptodactylidae)'. *Aust. J. Zool.* 21: 277–284.

Littlejohn, M.J. and Watson, G.F. (1974). 'Mating call discrimination and phonotaxis by females of the *Crinia laevis* complex (Anura: Leptodactylidae)'. *Copeia,* 1974(1): 171

Littlejohn, M.J. and Watson, G.F. (1976). 'Effectiveness of a hybrid mating call in eliciting phonotaxis by females of the *Geocrinia laevis* complex (Anura: Leptodactylidae)'. *Copeia,* 1976(1): 76–79.

Littlejohn, M.J. and Watson, G.F. (1983). 'The *Litoria ewingi* complex (Anura: Hylidae) in south–eastern Australia. VII. Mating–call structure and genetic compatability across a narrow hybrid zone between *L. ewingi* and *L. paraewingi'. Aust. J. Zool.* 31: 193–204.

Littlejohn, M.J. and Watson, G.F. (1985). 'Hybrid zones and homogamy'. *Ann. Rev. Ecol. Syst.* 16: 85–112.

Littlejohn, M.J., Watson, G.F. and Loftus–Hills, J.J. (1971). 'Contact hybridization in the *Crinia laevis* complex (Anura: Leptodactylidae)'. *Aust.J. Zool.* 19: 85–100.

Loftus–Hills, J.J. and Johnstone, B.M. (1970). Auditory function, communication and the brain–evoked response in anuran amphibians. *J. Acovst. Soc. Amer.* 47: 1131–1138.

Loftus–Hills, J.J. and Littlejohn, M.J. (1971a), 'Pulse repetition rate for mating-call discrimination by two sympatric species of *Hyla'. Copeia* 1971(1): 154–156.

Loftus–Hills, J.J. and Littlejohn, M.J. (1971b). 'Mating–call sound intensities of anuran amphibians'. *J. Acovst. Soc. Amer.* 49(4): 1327–1329.

Ludemann, D. and Neumann, H. (1961).

'Versuche uber die akute toxische wirkung neuzeitlicher kontaklinsectizide auf sub-wassertiere (Z. Beitrag)'. *Zeitsch. Angew. Zool.* 1960–61: 303–321.

Lynch, J.D. (1971). 'Evolutionary relationships, osteology, and zoogeography of leptodactyloid frogs'. *Misc. Publ. Univ. Kansas Mus. Nat. Hist.* (53): 1–238.

Lynch, J.D. (1973). 'The transition from archaic to advanced frogs'. *In* J.L. Vial (Ed.) *Evolutionary biology of the anurans. Contemporary research on major problems.* Univ. Missouri Press, Columbia.

MacNally, R.C. (1983). 'Trophic relationships of two sympatric species of *Ranidella* (Anura)'. *Herpetologica,* 39(2): 130–140.

Mahony, M.J. and Roberts, J.D. (1986). 'Two new species of desert burrowing frogs of the genus *Neobatrachus* (Anura: Myobatrachidae) from Western Australia'. *Rec. West. Aust. Mus.* 13(1): 155–170.

Mahony, M., Tyler, M.J. and Davies, M. (1984). 'A new species of the genus *Rheobatrachus* (Anura: Leptodactylidae) from Queensland'. *Trans. R. Soc. S. Aust.* 108, 155–162.

Main, A.R. (1957). 'Studies on Australian Amphibia. 1. The genus *Crinia* Tschudi in south–western Australia, and some species from south–eastern Australia'. *Aust. J. Zool.* 5: 30–55.

Main, A.R. (1964). 'A new species of *Pseudophyrne* (Anura: Leptodactylidae) from north–western Australia'. *West. Aust. Nat.* 9: 66–72.

Main, A.R. (1965). *Frogs of southern Western Australia.* Western Australian Naturalists' Club, Perth.

Main, A.R. (1968). 'Ecology, systematics and evolution of Australian frogs'. *In* J.B. Cragg (Ed.) *Advances in ecological research,* 5: 37–86.

Main, A.R. and Calaby, J.H. (1957). 'New records and notes on the biology of frogs from north–western Australia'. *West. Aust. Nat.* 5: 216–228.

Mallone, B.S. (1985). 'Mortality during the early life history stages of the Baw Baw frog. *Philoriafrosti* (Anura: Myobatrachidae)'. *In* G. Grigg, R. Shine and H. Ehmann (Eds) *Biology of Australasian frogs and reptiles.* Surrey Beatty & Sons, Chipping Norton, NSW.

Marchal–Segault, D. (1976). 'Toxicite de quelques insecticides pour des tetards de *Bufo bufo* (Amphibiens, Anoures)'. *Bull. Ecol.* 7: 411–416.

Martin, A.A. (1965). 'Tadpoles of the Melbourne area'. *Vic. Nat.* 8: 139–149.

Martin, A.A. (1967a). 'Australian anuran life histories: some evolutionary and ecological aspects'. *In* A.H. Weatherley (Ed.) *Australian Inland Waters and their fauna.* Australian National University, Canberra.

Martin, A.A. (1967b). 'The early development of Tasmania's endemic Anura, with comments on their relationships'. *Proc. Linn. Soc. NSW,* 92: 107–116.

Martin, A.A. (1970). 'Parallel evolution in the adaptive ecology of leptodactylid frogs of South America and Australia'. *Evolution,* 24: 643–644.

Martin, A.A. (1972). 'Studies in Australian Amphibia. 111. The *Limnodynastes dorsalis* complex (Anura: Leptodactylidae)'. *Aust. J. Zool.* 20: 165–211.

Martin, A.A. and Littlejohn, M.J. (1966). 'The breeding biology and larval development of *Hyla jervisiensis* (Anura: Hylidae)'. *Proc. Linn. Soc. NSW,* 91(1): 47–57.

Martin, A.A. and Littlejohn, M.J. (1982). *Tasmanian Amphibians.* Fauna of Tasmania handbook No. 6. University of Tasmania, Hobart.

Martin, A.A., Littlejohn, M.J. and Rawlinson, P.A. (1966). 'A key to anuran eggs of the Melbourne area, and an addition to the anuran fauna'. *Vict. Nat.* 83: 312–315.

Martin, A.A. and Tyler, M.J. (1978). 'The introduction into Western Australia of the frog *Limnodynastes tasmaniensis*'. *Aust. Zool.* 19: 320–324.

Martin, A.A., Tyler, M.J. and Davies, M. (1980). 'A new species of *Ranidella* (Anura: Leptodactylidae) from northwestern Australia'. *Copeia,* 1980: 93–99.

Martin, A.A. and Watson, G.F. (1971). 'Life history as an aid to generic delimitation in the family Hylidae'. *Copeia,* 1971: 78–89.

Martin, A.A., Watson, G.F., Gartside, G.F., Littlejohn, M.J. and Loftus–Hills, J.J. (1979). 'A new species of the *Litoria peronii* complex (Anura: Hylidae) from eastern Australia'. *Proc. Linn. Soc. NSW,* 103: 23–35.

Mathur, D.S. and Rane, P.D. (1979). 'Histopathological changes in the liver and intestine of *Rana cyanophlyctis* (Schn.) induced by endrin'. *Proc. Symp. Environ. Biol.*: 415–417. Academy of Environmental Biology, India.

Maxson, L.R. and Roberts, J.D. (1985). 'An

immunological analysis of the phylogenetic relationships between two enigmatic frogs. *Myobatrachus* and *Arenophryne*'. *J. Zool.* 207: 289–300.

Maxson, L.R., Tyler, M.J. and Maxson, R.D. (1982). 'Phylogenetic relationships of *Cyclorana* and the *Litoria aurea* species–group (Anura: Hylidae): a molecular perspective'. *Aust. J. Zool.* 30: 643–651.

McDonald, K.R. and Miller, J.D. (1982). 'On the status of *Lechriodus fletcheri* (Boulenger) (Anura: Leptodactylidae) in Northeast Queensland'. *Trans. R. Soc. S. Aust.* 106(4): 220.

McDonald, K.R. & Davies, M. (1990). 'Morphology and biology of the Australian tree frog *Litoria pearsoniana* (Copland) (Anura: Hylidae). *Trans. R. Soc. S. Aust.* 114(3): 145–156.

McDonald, K.R. and Tyler, M.J. (1984). 'Evidence of gastric brooding in the Australian leptodactylid frog *Rheobatrachus vitellinus*'. *Trans. R. Soc. S. Aust.* 108: 226.

McDonnell, L.J., Gartside, D.F. and Littlejohn, M.J. (1978). 'Analysis of a narrow hybrid zone between two species of *Pseudophryne* (Anura: Leptodactylidae) in south–eastern Australia'. *Evolution*, 32(3): 602–612.

Meeks, R.L. (1968). 'The accumulation of 36C1 ring–labelled DDT in a freshwater marsh'. *J. Wildl. Man.* 32: 376–398.

Menzies, J.I. (1987). 'A taxonomic revision of Papuan *Rana*'. *Aust. J. Zool.* 35: 373–418.

Mohanty–Hejmadi, P. and Dutta, S.K. (1981). 'Effects of some pesticides on the development of the Indian bull frog *Rana tigerina*'. *Environ. Pollut.* (Ser. A), 24: 145–161.

Moore, J.A. (1958). 'A new genus and species of leptodactylid frog from Australia'. *Amer. Mus. Novit.* (1919): 1–7.

Moore, J.A. (1961). 'Frogs of eastern New South Wales'. *Bull. Amer. Mus. Nat. Hist.* 121: 149–386.

Morisson, J.M. (1982). *Gas exchange in amphibian eggs*. B.Sc.(Hons) thesis (Unpublished), Dept of Zoology, University of Adelaide.

Nishimura, K. (1967). 'Abnormal formation of visual organs of amphibian larvae induced by radioactive rainwater'. *Mie Med. J.* 16: 263–267.

Niven, B.S. and Stewart, M.G. (1982). 'The precise environment of some well–known animals; VII, the Cane Toad *(Bufo marinus)*'. *AES Working Paper* 3/82, Griffith University, School

of Environmental Studies, Brisbane.

Odendaal, F.J. and Bull, C.M. (1980). 'Influence of water speed on tadpoles of *Ranidella signifera* and *R. riparia* (Anura: Leptodactylidae) from South Australia'. *Aust. J. Zool.* 28: 79–82.

Odendaal, F.J., Bull, C.M. and Telford, S.R. (1983). The vocabulary of calls of *Ranidella riparia* (Anura: Leptodactylidae). *Copeia*, 1983(2): 534–537.

Odendaal, F.J., Bull, C.M. and Telford, S.R. (1986). Influence of the acoustic environment on the distribution of the frog *Ranidella riparia*. *Anim. Behav.* 34: 1836–1843.

Oliver, J.A. (1943). 'The peripatetic toad'. *Nat. Hist.* 58(1): 30–33.

Osborn, D. and French, M.C. (1981). 'The toxicity of the mothproofing chemical EULAN WA NEW to frog *Rana temporaria* tadpoles'. *Environ. Pollut.* (Ser. A), 24: 117–123.

Osborn, D., Cooke, A.S. and Freestone, S. (1981). 'Histology of a teratogenic effect of DDT on *Rana temporaria* tadpoles'. *Environ. Pollut.* (Ser. A) 25: 305–319.

Otani, A., Palumbo, N. and Read, G. (1969). 'Pharmacodynamics and treatment of mammals poisoned by *Bufo marinus* toxin'. *Amer. J. Vet. Res.* 30: 1865–1872.

Parker, H.W. (1940). 'Australasian frogs of the family Leptodactylidae'. *Novit. Zool.* 42(1): 1–106.

Pasanan, S. and Koskela, P. (1976). 'Seasonal variations in the mineral content of certain organs in the common frog, *Rana temporaria* L'. *Joensuun Korkeakoulun Julkaisuja*, (Ser. B): l–7.

Pasteels, J. (1942). 'Les effets du LiCl sur le developpement de *Ranafusca*'. *Bull. Acad. Roy. Belg. Cl. Sci.* 5th Ser., 28: 605–614.

Pasteels, J. (1945). 'Recherches sur l'action du LiCl sur les oeufs des amphibiens'. *Archiv. Biol.* 56: 105–183.

Paulov, S. (1976). 'Evidence for inhibition by juveniles of the development in vertebrates'. *Acta Vet. Brno*, 45: 215–220.

Paulov, S. (1977a). 'Effect of the herbicide Gramoxone S (dichloride paraquat) on the development and body proteins of amphibia *(Rana temporaria* L.)'. *Biologia (Bratislava)*, 32: 127131.

Paulov, S. (1977b). 'The effects of N–phenyl–N'–methyl urea (Defenuron) on the development of amphibia *(Rana temporaria* L.)'. *Biologia (Bratislava)*, 32: 607–611.

Pemberton, C.E. (1933). 'Introduction to Hawaii of the tropical American Toad *Bufo marinus*'. *Hawaiian Planters' Rec*. 37: 15–16.

Pemberton, C.E. (1934). 'Local investigations on the introduced tropical American Toad *Bufo marinus*. *Hawaiian Planters' Rec*. 38: 186–192.

Pemberton, C.E. (1949). 'Longevity of the tropical American toad, *Bufo marinus*'. *Science,* 110: 512.

Pengilley, R.K. (1971a). 'The food of some Australian anurans (Amphibia)'. *J. Zool., Lond*. 163: 93–103.

Pengilley, R.K. (1971b). 'Calling and associated behaviour of some species of *Pseudophryne* (Anura: Leptodactylidae)'. *J. Zool., Lond*. 163: 73–92.

Pengilley, R.K. (1973). 'Breeding biology of some species of *Pseudophryne* (Anura: Leptodactylidae) of the southern highlands, New South Wales'. *Aust. Zool*. 18: 15–30.

Perez–Coll, C.S., Herkovits, J. and Salibian, A. (1985). 'Effects of cadmium on the development of an amphibian'. *Arch. Biol. Med. Exp.* Chile 18: 33–40.

Pippet, J.R. (1975). 'The Marine Toad, *Bufo marinus,* in Papua New Guinea'. *Papua New Guinea Agric. J*. 26(1): 23–30.

Pope, C.H. (1955). *The Reptile World*. A. Knopf, New York.

Puscariu, F., Papillian, V.V., Gabor, M. and Deac, C. (1973). 'Toxicite et effets morphopathologiques des aflatoxines chez les tetards'. *Arch. Roum. Path. exp. Microbiol*. 32: 255–267.

Rabor, D.S. (1952). 'Preliminary notes on the giant toad. *Bufo marinus* (Linn.), in the Philippine Islands'. *Copeia,* 1952: 281–282.

Rane, P.D. and Mathur, D.S. (1978). 'Toxicity of Aldrin to *Rana cyanophlyctis* (Schn)'. *Sci. Cult*. 44: 128–130.

Ranke–Rybicka, B. (1972). 'An inquiry into viability of tadpoles of *Rana temporaria* intermittently exposed to phosphoorganic pesticides (Foschlorine, Malathion)'. *Roczn. Pzn,* 23: 371–377.

Reimer, W.J. (1959). 'Giant toads of Florida'. *J. Fla Acad. Sci*. 21: 207–211.

Richards, C.M. (1962). 'The control of tadpole growth by Alga–like cells'. *Physiol. Zool*. 35(4): 285–296.

Richards, S.J. (1992). 'The tadpole of the Australian frog *Litoria nyakalensis* (Anura: Hylidae) and a key to the torrent tadpoles of northern Queensland'. *Alytes* 10: 99–103.

Richards, S.J. (1992). 'The tadpole of the Australopapuan frog *Rana daemeli*'. *Memoirs of the Queensland Museum,* 32: 138.

Richards, S.J. (1993). 'Functional significance of nest construction by an Australian rainforest frog: a preliminary analysis'. *Memoirs of the Queensland Museum,* 34: 89–93.

Richards, S.J. and Alford, R.A. (1992). 'Nest construction by an Australian rainforest frog of the *Litoria lesueuri* complex (Anura: Hylidae)'. *Copeia,* 1992 (4): 1120–1123.

Richards, S.J. and Alford, R.A. (1993). 'The tadpoles of two Queensland frogs (Anura: Hylidae, Myobatrachidae)'. *Memoirs of the Queensland Museum,* 33: 337–340.

Richards, S.J. and James, C. (1992). 'Foot–flagging displays of some Australian frogs'. *Memoirs of the Queensland Museum,* 132: 302.

Richards, S.J., McDonald, K.R. and Alford, R.A. (1993) 'Declines in populations of Australia's endemic tropical rainforest frogs'. *Pacific Conservation Biology,* 1: 66–77.

Roberts, J.D. (1978). 'Redefinition of the Australian leptodactylid frog *Neobatrachus pictus* Peters'. *Trans. R. Soc. S. Aust*. 102(4): 97–105.

Roberts, J.D. (1981). 'Terrestrial breeding in the Australian leptodactylid frog *Myobatrachus gouldii* (Gray)'. *Aust. Wildl. Res*. 8: 451–462.

Roberts, J.D. (1984). 'Terrestrial egg deposition and direct development in *Arenophryne rotunda* Tyler, a myobatrachid frog from coastal sand dunes at Shark Bay, WA'. *Aust. Wildl. Res*. 11: 191–200.

Roberts, J.D. (1985). 'Population estimates for *Arenophryne rotunda*: is the round frog rare?' *In* G. Grigg, R. Shine and H. Ehmann (Eds) *Biology of Australasian frogs and reptiles*. Surrey Beatty & Sons, Chipping Norton, NSW.

Roberts, J.D. (1990). 'The biology of *Arenophryne rotunda* (Anura: Myobatrachidae): a burrowing frog from Shark Bay, Western Australia'. pp 287–297 *In* P.F. Berry, S.D. Bradshaw and B.R. Wilson (Eds) Research in Shark Bay. *Report of the France–Austral Bicentenary Expedition Committee.*

Roberts, J.D., Wardell–Johnson, G. and Barendse, W. (1990). 'Extended descriptions of two new species of *Geocrinia* (Anura: Myobatrachidae) from south–western Australia, with comments on the status of *G. lutea*'. *Rec. West. Aust. Mus.* 14: 427–437.

Robertson, J.G.M. (1984). 'Acoustic spacing by breeding males of *Uperoleia rugosa* (Anura: Leptodactylidae)'. *Z. Tierpsychol.* 64: 283-297.

Robertson, J.G.M. (in press). 'The breeding behaviour of *Uperoleia rugosa* (Anura: Leptodactylidae): II female choice, male strategies and the role of vocalisations'. *Anim. Behav.*

Robinson, R.L. and Tyler, M.J. (1972). 'The catecholamine content of the adrenal glands of frogs as an index of phylogenetic relationships'. *Comp. Gen. Pharmacol.* 3(10): 167-170.

Ros, B.A. (1974). 'Gut contents of some amphibians and reptiles'. *Herpetofauna,* 7(1):4-8.

Rostand, J. (1957). 'Grenouilles monstreuses et radioactivite'. *Comptes Rendus. Acad. Sci.* 245: 1175-1176.

Rostand, J. (1958). *Anomalies des Ampllibiens Anoures.* Sedes, Paris.

Rzehak, K., Maryanska-Nadachowska, A. and Jordan, M. (1977). 'The effect of Karbatox 75, a carbaryl insecticide, upon the development of tadpoles of *Rana temporaria* and *Xenopus laevis*'. *Folia Biol. Krakow,* 25: 391-399.

Sabath, M.D., Boughton, W.C. and Easteal, S. (1981). 'Expansion of the range of the introduced toad *Bufo snarinus* in Australia from 1935 to 1974'. *Copeia,* 1981(3): 676-680.

Sanders, J. and Davies, M. (1984). 'Burrowing behaviour and associated hindlimb myology in some Australian hylid and leptodactylid frogs'. *Aust. Zool.* 21(2): 123-142.

Sanders, H.O. (197()). 'Pesticide toxicities to tadpoles of the Western chorus frog *Pseudacris triseriata* and Fowler's toad *Bufo woodhousiifowleri*'. *Copeia,* 1970: 246-251.

Savage, J.M. (1973). 'The geographic distribution of frogs: patterns and predictions'. *In* J.L. Vial (Ed.) *Evolutionary biology of the Anurans.* University of Missouri Press, Columbia.

Savage, J.M. (1986). 'Nomenclatural notes on the Anura (Amphibia)'. *Proc. biol. Soc. Wash.* 99(1): 42-45.

Schowing, J. and Boverio, A. (1979). 'Influence teratogene du bichlorure de mercure sur le development embryonnaire de l'amphibien *Xenopus laevis* Daud'. *Acta Embryol. Exp.* 1: 39-52.

Scorgie, H.R.A. and Cooke, A.S. (1979). 'Effects of the triazine herbicide Cynatryn on aquatic animals'. *Bull. Environ. Contam. Toxicol.* 22: 135-142.

Seymour, R.S. and Lee, A.K. (1974). 'Physiological adaptations of anuran amphibians to aridity: Australian prospects'. *Aust. Zool.* 18(2): 53-65.

Smyth, E.G. (1916). 'The white grubs injuring sugar cane in Porto Rico'. *Puerto Rico Univ. J. Agric.* 1: 141-169.

Sokol, A. (1974). 'Plasticity in the fine timing of metamorphosis in tadpoles of the hylid frog, *Litoria ewingi*'. *Copeia,* 1984: 868-873.

Spencer, B. (1896). *Report on the work of the Horn Expedition to Central Australia.* Dulau, London.

Stammer, D. (1981). 'Some notes on the cane toad *(Bufo marinus)*'. *Aust. J. Herpetol.* 1(2): 61.

Stanton, J.P. and Morgan, M.G. (1977). *Project 'Rakes'—a rapid appraisal of key and endangered sites. Report No. 1: The rapid selection of key and endangered sites: the Queensland case study.* Univ. New England, School of Natural Resources, Armidale, NSW. Mimeographed.

Stille, W.T. (1958). 'The water absorption response of an anuran'. *Copeia,* 1958: 217-218.

Straughan, I.R. (1969). 'The *Hyla bicolor* complex (Anura: Hylidae) in North Queensland'. *Proc. R. Soc. Qld.* 80(5): 43-54.

Straughan, I.R. and Lee, A.K. (1966). 'A new genus and species of leptodactylid frog from Queensland'. *Proc. R. Soc. Qld.* 77: 63-66.

Straughan, I.R. and Main, A.R. (1966). 'Speciation and polymorphism in the genus *Crinia* Tschudi (Anura: Leptodactylidae) in Queensland'. *Proc. R. Soc. Qld.* 78: 11-28.

Stromme, J.E., Maggert, J.E. and Scholander, P.F. (1969). 'Interstitial fluid pressure in terrestrial and semiterrestrial animals'. *J. Appl. Physiol.* 27(1): 123-126.

Stuart, L.C. (1951). 'The distributional implications of temperature tolerances and haemoglobin values in the toads *Bufo marinus* (Linnaeus) and *Bufo bocourti* Brocchi'. *Copeia,* 1951 (3): 220-228.

Taylor, P.M., Tyler, M.J. and Shearman, D.J.C. (1985a). 'Gastric acid secretion in the toad *Bufo marinus* with the description of a new technique for *in vivo* monitoring'. *Comp. Biochem. Physiol.* 81A(2): 325-327.

Taylor, P.M., Tyler, M.J. and Shearman, D.J.C. (1985b). 'Gastric emptying and intestinal transit in *Bufo marinus* and the action of E prostaglandins'. *Aust. J. Exp. Biol. Med. Sci.* 63: 223-230.

Tyler, M.J. (1971a). 'The phylogenetic significance of vocal sac structure in hylid frogs'. *Univ.*

Kansas Publ. Mus. Nat. Hist. 19(4): 319-360.

Tyler, M.J. (1971b). 'Voluntary control of the shape of the inflated vocal sac by the Australian leptodactylid frog *Limnodynastes tassnaniensis*'. *Trans. R. Soc. S. Aust.* 95(1): 49-52.

Tyler, M.J. (1972). 'A new genus for the Australian leptodactylid frog *Crinia darlingtoni*'. *Zool. Meded.* 47: 193-201.

Tyler, M.J. (1974). 'First frog fossils from Australia'. *Nature,* 248: 711-712.

Tyler, M.J. (1975). 'The ontogeny of the vocal sac of the Australian leptodactylid frog *Limnodynastes tasmaniensis*'. *Trans. R. Soc. S. Aust.* 99(2): 85-87.

Tyler, M.J. (1976a). 'Comparative osteology of the pelvic girdle of Australian frogs and description of a new fossil genus'. *Trans. R. Soc. S. Aust.* 100: 3-14.

Tyler, M.J. (1976b). *Frogs.* Collins, Sydney.

Tyler, M.J. (1976c). 'A new genus and two new species of leptodactylid frogs from Western Australia'. *Rec. West. Aust. Mus.* 4: 45-52.

Tyler, M.J. (1977). 'Pleistocene frogs from caves at Naracoorte, South Australia'. *Trans. R. Soc. S. Aust.* 101: 85-89.

Tyler, M.J. (1978). *Amphibians of South Australia.* Government Printer, Adelaide.

Tyler, M.J. (1979). 'Herpetofaunal relationships of South America with Australia'. *In* W.E. Duellman (Ed.) *The South American herpetofauna: its origin, evolution and dispersal.* Monogr. Mus. Nat. Hist. Univ. Kansas 7: 73-106.

Tyler, M.J. (1982). 'Tertiary frogs from South Australia'. *Alcheringa,* 6: 101-103.

Tyler, M.J. (Ed.) (1983a). *The gastric brooding frog.* Croom Helm, London & Canberra.

Tyler, M.J. (1985b). 'Miscellany'. *In* M.J. Tyler (Ed.) *The gastric brooding frog.* Croom Helm, London & Canberra.

Tyler, M.J. (1983c). 'Oral birth and perinatal behaviour'. *In* M.J. Tyler (Ed.) *The gastric brooding frog.* Croom Helm, London & Canberra.

Tyler, M.J. (1985a). 'Quaternary fossil frogs from Skull Cave and Devil's Lair in the extreme south-west of Western Australia'. *Rec. W. Aust. Mus.* 12: 233-240.

Tyler, M.J. (1985b). 'Nomenclature of the Australian herpetofauna: anarchy rules O.K.'. *Herpetol. Rev.* 16(3): 69.

Tyler, M.J. (1985c). 'A crisis in zoological nomenclature'. *Search,* 16(9012): 237.

Tyler, M.J. (1985d). *There's a frog in my (throat) stomach.* Collins, Sydney.

Tyler, M.J. (1986). 'Additional records of *Australobatrachus ilius* (Anura: Hylidae) from the Tertiary of South Australia'. *Alcheringa,* 8: 401-402.

Tyler, M.J. (1988). '*Neobatrachus pictus* (Anura: Leptodactylidae) from the Miocene/Pliocene boundary of South Australia'. *Trans. R. Soc. S. Aust.* 112.

Tyler, M.J. (1989). 'A new species of *Lechriodus* (Anura: Leptodactylidae) from the Tertiary of Queensland, with a redefinition of the ilial characteristics of the genus'. *Trans. R. Soc. S. Aust.* 113, 15-21.

Tyler, M.J. (1990). '*Limnodynastes* Fitzinger (Anura: Leptodactylidae) from the Cainozoic of Queensland'. *Mem. Qld. Mus.* 29 (2).

Tyler, M.J. (1991a). '*Crinia* Tschudi (Anura: Leptodactylidae) from the Cainozoic of Queensland, with the description of a new species'. *Trans. R. Soc. S. Aust.* 115(2), 99-101.

Tyler, M.J. (1991b). '*Kyarranus* Moore (Anura: Leptodactylidae) from the Tertiary of Queensland'. *Proc. R. Soc. Vic.* 103(1), 47-51.

Tyler, M.J. (1991c). 'A large new species of *Litoria* (Anura: Hylidae) from the Tertiary of Queensland'. *Trans. R. Soc. S. Aust.* 115(2), 103-105.

Tyler, M.J. (1991d). 'Declining amphibian populations—a global phenomenon? An Australian perspective'. *Alytes.* 9(2), 43-50.

Tyler, M.J. (1991e). 'Where have all the frogs gone?'. *Aust. Nat. Hist.* 23(8), 618-625.

Tyler, M.J. (1994). 'The Action Plan for Australian Frogs'. *Australian National Conservation Agency,* Canberra.

Tyler, M.J. (In press). 'Hylid frogs from the Mid-Miocene Camfield Beds of northern Australia'. *Beagle.*

Tyler, M.J. and Anstis, M. (1975). 'Taxonomy and biology of frogs of the *Litoria citropa* complex (Anura: Hylidae)'. *Rec. S. Aust. Mus.* 17: 41-50.

Tyler, M.J., Aslin, F.W. and Bryars, S. (1992). 'Early Holocene frogs from the Tantanoola Caves, South Australia'. *Trans. R. Soc. S. Aust.* 116, 153.

Tyler, M.J. and Cappo, M.C. (1983). 'Diet and feeding habits of frogs of the Magela Creek System (Final Report)'. *Open File Record 10,* Supervising Scientist for the Alligator Rivers Region, Sydney.

Tyler, M.J., Crook, G.A. and Davies, M. (1983). 'Reproductive biology of the frogs of the Magela Creek System, Northern Territory'. *Rec. S. Aust. Mus.* 18: 415-440.

Tyler, M.J. and Davies, M. (1978). 'Species-groups within the Australopapuan hylid frog genus *Litoria* Tschudi'. *Aust. J. Zool.* Suppl. Ser. (63): 1-47.

Tyler, M.J. and Davies, M. (1979). 'Foam nest construction by Australian leptodactylid frogs (Amphibia, Anura, Leptodactylidae). *J. Herpetol.* 13(4): 509-510.

Tyler, M.J. and Davies, M. (1983). 'Larval development'. *In* M.J. Tyler (Ed.) *The gastric brooding frog*. Croom Helm, London & Canberra.

Tyler, M.J. and Davies, M. (1986). *Frogs of the Northern Territory*. Conservation Commission of the Northern Territory, Alice Springs.

Tyler, M.J., Davies, M. and Martin, A.A. (1978). 'A new species of hylid frog from the Northern Territory'. *Trans. R. Soc. S. Aust.* 106: 151-157.

Tyler, M.J., Davies, M. and Martin, A.A. (1981a). 'Australian frogs of the leptodactylid genus *Uperoleia* Gray'. *Aust. J. Zool.* Suppl. Ser. (79): 1-64.

Tyler, M.J., Davies, M. and Martin, A.A. (1981b). 'New and rediscovered species of frogs from the Derby-Broome area of Western Australia'. *Rec. West. Aust. Mus.* 9(2): 147-172.

Tyler, M.J., Davies, M. and Martin, A.A. (1981c). 'Frog fauna of the Northern Territory: new distributional records and the description of a new species'. *Trans. R. Soc. S. Aust.* 105(3): 149-154.

Tyler, M.J., Davies, M. and Martin, A.A. (1983). 'The frog fauna of the Barkly Tableland, Northern Territory'. *Trans. R. Soc. S. Aust.* 107(4): 237-242.

Tyler, M.J., Davies, M. and Walker, K.F. (1985). 'Abrasion injuries in burrowing frogs (Amphibia: Anura) from the Northern Territory, Australia'. *Zool. Anz.* 214(1-2): 54-60.

Tyler, M.J., Davies, M. and Watson, G.F. (1987) 'Frogs of the Gibb River Road, Kimberley Division, Western Australia. *Rec. W. Aust. Mus.* 13(3): 541-552.

Tyler, M.J. and Martin, A.A. (1975). 'Australian frogs of the *Cyclorana australis* complex'. *Trans. R. Soc. S. Aust.* 99(2): 93-99.

Tyler, M.J., Martin, A.A. and Davies, M. (1979). 'Biology and systematics of a new limnodynastine genus (Anura: Leptodactylidae) from

north-western Australia'. *Aust. J. Zool* 27: 135-150.

Tyler, M.J. and Parker, F. (1974). 'New species of hylid and leptodactylid frogs from southern New Guinea'. *Trans. R. Soc. S. Aust.* 98(2): 71-77.

Tyler, M.J., Roberts, J.D. and Davies, M. (1980). 'Field observations on *Arenophryne rotunda* Tyler, a leptodactylid frog of coastal sandhills'. *Aust. Wildl. Res.* 7: 295-304.

Tyler, M.J., Smith, L.A. and Johnstone, R.E. (1994). Second edition. *Frogs of Western Australia*. Western Australian Museum, Perth.

Tyler, M.J. and Watson, G.F. (1986). 'On the nomenclature of a hylid frog from Queensland'. *Trans. R. Soc. S. Aust.* 110(4): 193-194.

Tyler, M.J., Watson, G.F. and Davies, M. (1983). 'Additions to the frog fauna of the Northern Territory'. *Trans. R. Soc. S. Aust.* 107(4): 243-245.

Tyler, M.J., Hand, S.J. and Ward, V.J. (1990). 'Analysis of the frequency of *Lechriodus intergerivus* Tyler (Anura: Leptodactylidae) in Oligo-Miocene local faunas of Riversleigh Station, Queensland'. *Proc. Linn. Soc. N.S.W.* 112(2), 105-109.

Tyler, M.J., Davies, M. and Watson, G.F. (1991). 'The frog fauna of Melville Island, Northern Territory'. *The Beagle, Records of the Northern Territory Museum of Arts and Sciences,* 8(1):1-10.

Tyler, M.J. and Godthelp, H. (1993). 'A new species of *Lechriodus* Boulenger (Anura: Leptodactylidae) from the Early Eocene of Queensland'. *Trans. R. Soc. S. Aust.* 117, 187-189.

Tyler, M.J., Godthelp, H. and Archer, M. (in press). 'Frogs from a Plio-Pleistocene site at Floraville Station, northwest Queensland'. *Rec. S. Aust. Mus.*

Van Beurden, E. (1979). 'Gamete development in relation to season, moisture, energy reserve, and size in the Australian water-holding frog *Cyclorana platycephalus*'. *Herpetologica,* 35(4): 370-374.

Van Beurden, E. (1980a). *Report on the results of stage 3 of an ecological and physiological study of the Queensland Cane Toad* Bufo marinus. Report to Australian National Parks and Wildlife Service, Canberra. Mimeographed.

Van Beurden, E. (1980b). 'Mosquitoes *(Mimomyia elegans* (Taylor)) feeding on the introduced toad *Bufo marinus* (Linnaeus): implications for control of a toad pest'. *Aust. Zool.* 20(3): 501-504.

Van Beurden, E. (1984). 'Survival strategies of the Australian water-holding frog, *Cyclorana platycephalus*'. *In* H.G. Cogger and E.E. Cameron (Eds) *Arid Australia*. Australian Museum. Sydney.

Van Beurden, E.K. and Grigg, G.C. (1980). 'An isolated and expanding population of the introduced toad *Bufo marinus* in New South Wales'. *Aust. Wildl. Res.* 7: 305-310.

Van Beurden, E. and McDonald, K.R. (1980). 'A new species of *Cyclorana (Anura*: Hylidae) from northern Queensland'. *Trans. R. Soc. S. Aust.* 104(6): 193-195.

Van Valen, L. (1974). 'A natural model for the origin of some higher taxa'. *J. Herpetol.* 8: 109-121.

Waite, [E.R.] (1904). *In Notes and exhibits*. Proc. Linn. Soc. NSW 29: 557.

Waite, E.R. (1929). *The reptiles and amphibians of South Australia*. Government Printer, Adelaide.

Wassersug, R.J. and Karmazyn, M. (1984). '*Rana* tadpoles secrete prostaglandin E_2; but what is its role in anuran development?'. *Amer. Zool.* 24: 54A.

Waterhouse, D.F. (1974). 'The biological control of dung'. *Sci. Amer.* 230: 101-109.

Watson, G.F., Littlejohn, M.J., Gartside, D.F. and LoftusHills, J.J. (1985). 'The *Litoria ewingi* complex (Anura: Hylidae) in south-eastern Australia. VIII. Hybridization between *L. ewingi* and *L. verreauxi alpina* in the Mount Baw Baw area, south central Victoria'. *Aust. J. Zool.* 33: 143-152.

Watson, G.F., Loftus-Hills, J.J. and Littlejohn, M.J. (1971). 'The *Litoria ewingi* complex (Anura: Hylidae) in southeastern Australia'. *Aust. J. Zool.* 19: 401-416.

Watson, G.F. and Martin, A.A. (1973). 'Life history, larval morphology and relationships of Australian leptodactylid frogs'. *Trans. R. Soc. S. Aust.* 97: 33-45.

Watson, G.F. and Martin, A.A. (1979). 'Early development of the Australian green hylid frogs'. *Aust. Zool.* 20: 259-268.

Webb, G.A. (1983). Diet in a herpetofaunal community on the Hawkesbury Sandstone Formation in the Sydney area. *Herpetofauna*, 14(2): 87-91.

Webb, G.A. (1987). A note on the distribution and diet of the giant burrowing frog *Heleioporus australiacus*. (Shaw and Nodder 1795) (Anura: Myobatrachidae). *Herpetofauna*, 17(2): 20-21.

Weis, J.S. (1975). 'The effect of DDT on tail regeneration in *Rana pipiens* and R. *catesbeiana* tadpoles'. *Copeia*, 1975: 765-767.

Wells, R.W. and Wellington, C.R. (1984). 'A synopsis of the Class Reptilia in Australia'. *Aust. J. Herpetol.* 1(3-4): 73-129.

Wells, R.W. and Wellington, C.R. (1985). 'A classification of the Amphibia and Reptilia of Australia'. *Aust. J. Herpetol.*, Suppl. Ser. (1): 1-61.

White, A.W., Whitford, D. and Watson, G.F. (1980). 'Redescription of the Jervis Bay Tree Frog *Litoria jervisiensis* ((Anura: Hylidae), with notes on the identity of Krefft's frog *(Litoria kreffti))*'. *Aust. Zool.* 20(3): 375-390.

Williams, W.D. (1987). 'Limnology, the study of inland waters: a comment on perceptions of salt-lake studies, past and present'. *In* P. De Decker and W. D . Williams (Eds) *Limnology in Australia*. CSIRO, Melbourne & W. Junk, The Hague.

Wiltshire, D.J. and Bull, C.M. (1977). 'Potential competitive interactions between larvae of *Pseudophryne bibroni* and *P. semimarmorata* (Anura: Leptodactylidae)'. *Aust. J. Zool.* 25: 449-454.

Wingate, D.B. (1965). 'Terrestrial herpetofauna of Bermuda'. *Herpetologica*, 21(3): 202-218.

Withers, P.C., Hillman, S.S. and Drewes, R.C. (1984). 'Evaporative water loss and skin lipids of anuran amphibians'. *J. Exper. Zool.* 232 11-17.

Wohlgemuth, E. (1977). 'Toxicity of Endrin to some species of aquatic vertebrates'. *Acta Sci. Nat. Acad. Sci. Bohemoslovacae Brno.* 11: 1-38.

Wohlgemuth, E. and Trnkova, J. (1978). 'Dispersal and effects of a preparation containing Endrin in an artificial water reservoir'. *Folia Zoologica* 28(1): 65-72.

Wojcik, J. and Ranke-Rybicka, B. (1971). 'The sensitivity to Tritox-30 (DDT, Gamma-HCH, DMTD) of the tadpoles of *Xenopus laevis* Daudin and *Rana temporaria* L.'. *Patstowy Zaklad Higeny Roczniki, 22*: 413-419.

Woodruff, D.S. (1976). 'Courtship, reproductive rates and mating system in three Australian *Pseudophryne* (Amphibia, Anura, Leptodactylidae)'. *J. Herpetol.* 10: 313-318.

Wotherspoon, D. (1981). 'The great barred frogmouth breeder?' *Herpetofauna* 12(2): 30.

Zug, G.R., Lindgren, E. and Pippet, J.R. (1975). 'Distribution and ecology of the Marine Toad, *Bufo marinus,* in Papua New Guinea'. *Pac. Sci.* 29: 31-50.

Zug, G.R. and Zug, P.B. (1979). 'The marine toad, *Bufo marinus*: a natural history resume of native populations'. *Smithson. Contrib. Zool.* (284): 1-58.

Zweifel, R.G. (1972). 'A review of the frog genus *Lechriodus* (Leptodactylidae) of New Guinea and Australia'. *Amer. Mus. Novit.* (2507): 1-41.

Zweifel, R.G. (1985). 'Australian frogs of the family Microhylidae'. *Bull. Amer. Mus. Nat. Hist.* 182: 265-388.

Identification Guides

(Currently or shortly available)

Barker, J., Grigg, G.C. and Tyler, M.J. (in press). *A field guide to Australian frogs.* Second edition. Surrey Beatty & Sons, Chipping Norton, NSW.

Cogger, H.G. (1992). *Reptiles and amphibians of Australia.* Revised edition. Reed, Sydney.

Hero, J.-M., Littlejohn, M. and Marantelli, G. (1991). *Frogwatch field guide to Victorian frogs.* Department of Conservation & Environment, East Melbourne.

Martin, A.A. and Littlejohn, M.J. (1982). *Tasmanian amphibians. Fauna of Tasmania handbook No. 6.* University of Tasmania, Hobart.

Robinson, M. (1993). *A field guide to frogs of Australia.* Reed, Sydney.

Tyler, M.J. (1977). *Frogs of South Australia.* Second edition. South Australian Museum, Adelaide.

Tyler, M.J. (1978). *Amphibians of South Australia.* Handbooks Committee, Adelaide.

Tyler, M.J. and Davies, M. (1986). *Frogs of the Northern Territory.* Conservation Commission of the Northern Territory, Alice Springs.

Tyler, M.J., Smith, L.A. and Johnstone, R.E. (1994). *Frogs of Western Australia.* Revised edition. Western Australian Museum, Perth.

APPENDIX
State and Territory Checklists

NEW SOUTH WALES

Adelotus brevis
Assa darlingtoni
Bufo marinus
Crinia deserticola
Crinia haswelli
Crinia parinsignifera
Crinia signifera
Crinia sloanei
Crinia tinnula
Cyclorana brevipes
Cyclorana novaehollandiae
Cyclorana platycephala
Cyclorana verrucosus
Heleioporus australiacus
Kyarranus loveridgei
Kyarranus sphagnicolus
Lechriodus fletcheri
Limnodynastes dumerilii
Limnodynastes fletcheri
Limnodynastes interioris
Limnodynastes ornatus
Limnodynastes peronii
Limnodynastes salmini
Limnodynastes tasmaniensis

Limnodynastes terraereginae
Litoria alboguttata
Litoria aurea
Litoria booroolongensis
Litoria brevipalmata
Litoria caerulea
Litoria chloris
Litoria citropa
Litoria dentata
Litoria ewingi
Litoria fallax
Litoria flavipunctata
Litoria freycineti
Litoria gracilenta
Litoria jervisiensis
Litoria latopalmata
Litoria lesueuri
Litoria nasuta
Litoria olongburensis
Litoria pearsoniana
Litoria peronii
Litoria phyllochroa
Litoria piperata
Litoria raniformis

Litoria rubella
Litoria spenceri
Litoria subglandulosa
Litoria tyleri
Litoria verreauxi
Mixophyes balbus
Mixophyes fasciolatus
Mixophyes fleayi
Mixophyes iteratus
Neobatrachus sudelli
Notaden bennetti
Pseudophryne australis
Pseudophryne bibroni
Pseudophryne coriacea
Pseudophryne corroboree
Pseudophryne dendyi
Taudactylus diurnus
Uperoleia capitulata
Uperoleia fusca
Uperoleia laevigata
Uperoleia martini
Uperoleia rugosa
Uperoleia tyleri

NORTHERN TERRITORY

Bufo marinus
Crinia bilingua
Crinia deserticola
Crinia remota
Cyclorana australis
Cyclorana cryptotis
Cyclorana cultripes
Cyclorana longipes
Cyclorana maculosus
Cyclorana maini
Cyclorana platycephala
Cyclorana vagitus
Limnodynastes
  convexiusculus
Limnodynastes ornatus
Limnodynastes spenceri
Limnodynastes tasmaniensis

Litoria alboguttata
Litoria bicolor
Litoria caerulea
Litoria coplandi
Litoria dahlii
Litoria inermis
Litoria infrafrenata
Litoria meiriana
Litoria microbelos
Litoria nasuta
Litoria pallida
Litoria personata
Litoria rothii
Litoria rubella
Litoria splendida
Litoria tornieri
Litoria wotjulumensis

Megistolotis lignarius
Neobatrachus aquilonius
Neobatrachus centralis
Neobatrachus sutor
Notaden melanoscaphus
Notaden nichollsi
Rana daemeli
Sphenophryne adelphe
Uperoleia arenicola
Uperoleia borealis
Uperoleia inundata
Uperoleia lithomoda
Uperoleia micromeles
Uperoleia orientalis
Uperoleia trachyderma

QUEENSLAND

Adelotus brevis
Assa darlingtoni
Bufo marinus
Cophixalus bombiens
Cophixalus concinnus
Cophixalus crepitans
Cophixalus exiguus
Cophixalus hosmeri
Cophixalus infacetus
Cophixalus mcdonaldi
Cophixalus neglectus
Cophixalus ornatus
Cophixalus peninsularis
Cophixalus saxatilis
Crinia deserticola
Crinia parinsignifera
Crinia remota
Crinia signifera
Crinia tinnula
Cyclorana australis
Cyclorana brevlpes
Cyclorana cultripes
Cyclorana maculosus
Cyclorana manya
Cyclorana novaehollandiae
Cyclorana platycephala
Cyclorana verrucosus
Kyarranus kundagungan
Kyarranus loveridgei
Lechriodus fletcheri
Limnodynastes convexiusculus
Limnodynastes dumerilii
Limnodynastes fletcheri
Limnodynastes ornatus
Limnodynastes peronii
Limnodynastes salmini
Limnodynastes spenceri
Limnodynastes tasmaniensis

Limnodynastes terraereginae
Litoria alboguttata
Litoria bicolor
Litoria brevipalmata
Litoria caerulea
Litoria chloris
Litoria cooloolensis
Litoria coplandi
Litoria dahlii
Litoria dentata
Litoria electricia
Litoria eucnemis
Litoria fallax
Litoria freycineti
Litoria genimaculata
Litoria gracilenta
Litoria inermis
Litoria infrafrenata
Litoria latopalmata
Litoria lesueuri
Litoria longirostris
Litoria lorica
Litoria microbelos
Litoria nannotis
Litoria nasuta
Litoria nigrofrenata
Litoria nyakalensis
Litoria olongburensis
Litoria pallida
Litoria peronii
Litoria pearsoniana
Litoria revelata
Litoria rheocola
Litoria rothii
Litoria rubella
Litoria subglandulosa
Litoria tyleri
Litoria verreauxi

Litoria wotjulumensis
Litoria xanthomera
Mixophyes fasciolatus
Mixophyes iteratus
Mixophyes schevilli
Mixophyes fleayi
Neobatrachus aquilonius
Neobatrachus sudelli
Notaden bennetti
Notaden melanoscaphus
Notaden nichollsi
Nyctimystes dayi
Pseudophryne bibroni
Pseudophryne coriacea
Pseudophryne major
Rana daemeli
Rheobatrachus silus
Rheobatrachus vitellinus
Sphenophryne fryi
Sphenophryne gracilipes
Sphenophryne pluvialis
Sphenophryne robusta
Taudactylus acutirostris
Taudactylus diurnus
Taudactylus eungellensis
Taudactylus liemi
Taudactylus pleione
Taudactylus rheophilus
Uperoleia altissima
Uperoleia capitulata
Uperoleia fusca
Uperoleia inundata
Uperoleia laevigata
Uperoleia lithomoda
Uperoleia littlejohni
Uperoleia mimula
Uperoleia rugosa
Uperoleia trachyderma

SOUTH AUSTRALIA

Crinia deserticola
Crinia parinsignifera
Crinia riparia
Crinia signifera
Cyclorana cultripes
Cyclorana maini
Cyclorana platycephala
Geocrinia laevis
Limnodynastes dumerilii
Limnodynastes fletcheri

Limnodynastes peronii
Limnodynastes spenceri
Limnodynastes tasmaniensis
Litoria caerulea
Litoria ewingi
Litoria latopalmata
Litoria peronii
Litoria raniformis
Litoria rubella
Neobatrachus centralis

Neobatrachus pictus
Neobatrachus sudelli
Neobatrachus sutor
Notaden nichollsi
Pseudophryne bibroni
Pseudophryne occidentalis
Pseudophryne semimarmorata
Uperoleia capitulata

TASMANIA

Crinia signifera
Crinia tasmaniensis
Geocrinia laevis
Limnodynastes dumerilii

Limnodynastes peronii
Limnodynastes tasmaniensis
Litoria burrowsae
Litoria ewingi

Litoria raniformis
*Pseudophryne
  semimarmorata*

VICTORIA

Crinia haswelli
Crinia parinsignifera
Crinia signifera
Crinia sloanei
Geocrinia laevis
Geocrinia victoriana
Heleioporus australiacus
Limnodynastes dumerilii
Limnodynastes fletcheri
Limnodynastes interioris
Limnodynastes peronii

Limnodynastes tasmaniensis
Litoria aurea
Litoria citropa
Litoria ewingi
Litoria jervisiensis
Litoria lesueuri
Litoria paraewingi
Litoria peronii
Litoria phyllochroa
Litoria raniformis
Litoria spenceri

Mixophyes balbus
Neobatrachus pictus
Neobatrachus sudelli
Philoria frosti
Pseudophryne bibroni
Pseudophryne dendyi
*Pseudophryne
  semimarmorata*
Uperoleia laevigata
Uperoleia martini
Uperoleia tyleri

WESTERN AUSTRALIA

Arenophryne rotunda
Crinia bilingua
Crinia georgiana
Crinia glauerti
Crinia insignifera
Crinia pseudinsignifera
Crinia subinsignifera
Cyclorana australis
Cyclorana cryptotis
Cyclorana cultripes
Cyclorana longipes
Cyclorana maini
Cyclorana platycephala
Cyclorana vagitus
Geocrinia alba
Geocrinia leai
Geocrinia lutea
Geocrinia rosea
Geocrinia vitellina
Heleioporus albopunctatus
Heleioporus barycragus
Heleioporus eyrei
Heleioporus inornatus
Heleioporus psammophilus
Limnodynastes convexiusculus
Limnodynastes depressus

Limnodynastes dorsalis
Limnodynastes ornatus
Limnodynastes spenceri
Limnodynastes tasmaniensis
Litoria adelaidensis
Litoria bicolor
Litoria caerulea
Litoria cavernicola
Litoria coplandi
Litoria cyclorhynchus
Litoria dahlii
Litoria inermis
Litoria meiriana
Litoria microbelos
Litoria moorei
Litoria nasuta
Litoria pallida
Litoria rothii
Litoria rubella
Litoria splendida
Litoria tornieri
Litoria wotjulumensis
Megistolotis lignarius
Myobatrachus gouldii
Neobatrachus albipes
Neobatrachus aquilonius

Neobatrachus centralis
Neobatrachus fulvus
Neobatrachus kunapalari
Neobatrachus pelobatoides
Neobatrachus sutor
Neobatrachus wilsmorei
Notaden melanoscaphus
Notaden nichollsi
Notaden weigeli
Pseudophryne douglasi
Pseudophryne guentheri
Pseudophryne nichollsi
Pseudophryne occidentalis
Uperoleia aspera
Uperoleia borealis
Uperoleia crassa
Uperoliea glandulosa
Uperoleia lithomoda
Uperoleia marmorata
Uperoleia micromeles
Uperoleia minima
Uperoleia mjobergi
Uperoleia russelli
Uperoleia talpa
Uperoleia trachyderma

AUSTRALIAN CAPITAL TERRITORY

Crinia parinsignifera
Crinia signifera
Limnodynastes dumerilii
Limnodynastes peroni
Limnodynastes tasmaniensis
Litoria aurea
Litoria flavipunctata

Litoria latopalmata
Litoria lesueuri
Litoria peroni
Litoria raniformis
Litoria spenceri
Litoria verreauxi alpina
Litoria verreauxi verreauxi

Neobatrachus sudelli
Pseudophryne bibroni
Pseudophryne corroboree
Pseudophryne dendyi
Uperoleia laevigata

INDEX